Cover image from GAO represents gaseous carbon dioxide (CO_2) molecules in ambient air, currently measured at around 390 parts per million. Carbon dioxide consists of a central carbon atom doubly bonded with two oxygen atoms ($O=C=O$). Carbon dioxide is colorless and odorless at room temperature. Plants consume CO_2 by photosynthesis, which converts CO_2 into nutrients using energy from the Sun. Many scientists believe that the increased atmospheric CO_2 concentration has increased the acidity of ocean waters and is the primary cause of increased global average surface temperature. Global management of CO_2 and related risks underlies current ideas about engineering the global climate system.

Highlights of GAO-11-71, a report
to the Ranking Member, Committee
on Science, Space, and Technology,
House of Representatives

Climate engineering
Technical status, future directions, and potential responses

Why GAO did this study

Reports of rising global temperatures have raised questions about responses to climate change, including efforts to (1) reduce carbon dioxide (CO_2) emissions, (2) adapt to climate change, and (3) design and develop climate engineering technologies for deliberate, large-scale intervention in Earth's climate.

Reporting earlier that the nation lacks a coordinated climate-change strategy that includes climate engineering, GAO now assesses climate engineering technologies, focusing on their technical status, future directions for research on them, and potential responses.

To perform this technology assessment, GAO reviewed the peer-reviewed scientific literature and government reports, consulted experts with a wide variety of backgrounds and viewpoints, and surveyed 1,006 adults across the United States. Experts convened with the assistance of the National Academy of Sciences advised GAO, and several reviewed a draft of this report. GAO incorporated their technical and other comments in the final report as appropriate.

View GAO-11-71 or key components at www.gao.gov. For more information, contact Timothy Persons at (202) 512-6412 or personst@gao.gov.

Report multimedia

Depiction of the global carbon cycle changes over time
www.gao.gov/multimedia/
interactive/GAO-11-71a

Global average energy budget of the Earth's atmosphere
www.gao.gov/multimedia/
interactive/GAO-11-71b

What GAO found

Climate engineering technologies do not now offer a viable response to global climate change. Experts advocating research to develop and evaluate the technologies believe that research on these technologies is urgently needed or would provide an insurance policy against worst case climate scenarios—but caution that the misuse of research could bring new risks. Government reports and the literature suggest that research progress will require not only technology studies but also efforts to improve climate models and data.

The technologies being proposed have been categorized as carbon dioxide removal (CDR) and solar radiation management (SRM). CDR would reduce the atmospheric concentration of CO_2, allowing more heat to escape and thus cooling the Earth. For example, proposed CDR technologies include enhancing the uptake of CO_2 in oceans and forests and capturing CO_2 from air chemically for storage underground. SRM technologies would place reflective material in space or in Earth's atmosphere to scatter or reflect sunlight (for example, by injecting sulfate aerosols into the stratosphere to scatter incoming solar radiation or brightening clouds) or would increase the planet's reflectivity (for example, by painting roofs and pavements in light colors). (See figure.)

Examples of climate engineering technologies. Source: GAO.

GAO found these technologies currently immature, many with potentially negative consequences. Some studies say, for example, that stratospheric aerosols might greatly reduce summer precipitation in places such as India and northern China.

Many experts advocated research because of its potential benefits but also recognized its risks. For example, a country might unilaterally deploy a technology with a transboundary effect. Research advocates emphasized the need for risk management, envisioning a federal research effort that would (1) focus internationally on transparency and cooperation, given transboundary effects; (2) enable the public and national leaders to consider issues before they become crises; and (3) anticipate opportunities and risks. A small number of those we consulted opposed research; they anticipated major technology risks or limited future climate change.

Based on GAO's survey, a majority of U.S. adults are not familiar with climate engineering. When given information on the technologies, they tend to be open to research but concerned about safety.

GAO
Accountability ★ Integrity ★ Reliability

July 28, 2011

The Honorable Eddie Bernice Johnson
Ranking Member
Committee on Science, Space, and Technology
House of Representatives

Dear Ms. Johnson:

In response to committee reports accompanying the legislative branch fiscal year 2008 appropriations bill, the U.S. Government Accountability Office established a permanent operational technology assessment group within GAO's Applied Research and Methods team: the Center for Science, Technology, and Engineering. Responding to your request that we conduct a technology assessment on proposed technological approaches toward engineering the climate, we examined the current state of climate engineering science and technology, experts' views of the future of U.S. climate engineering research, and potential public responses to climate engineering. We also discuss in this report key considerations for the use of climate engineering technologies and their policy implications.

As agreed with your office, we plan no distribution of this report until 14 days after its issue date unless you publicly announce its contents earlier. We will then send copies of this report to interested congressional committees; the Secretaries of Agriculture, Commerce, Defense, Energy, and State; the Administrator of the National Aeronautics and Space Administration; the Administrator of the Environmental Protection Agency; and the Director of the National Science Foundation. We will provide copies to others on request. In addition, the report will be available at no charge on the GAO website at www.gao.gov.

If you have any questions concerning this report, you may contact me at (202) 512-6412 or personst@gao.gov. Contact points for our Offices of Congressional Relations and Public Affairs may be found on the last page of this report. Major contributors to this report are listed on page 117.

Sincerely yours,

T. M. Persons

Timothy M. Persons, Ph.D.
Chief Scientist

Summary

Reports of rising global average surface temperature have raised interest in the potential for engineering Earth's climate, supplementary to ongoing efforts to reduce greenhouse gas emissions and prepare for climate change through adaptation. Proposed climate engineering technologies, or direct, deliberate, large-scale interventions in Earth's climate, generally aim at either carbon dioxide removal (CDR) or solar radiation management (SRM). Whereas CDR would reduce the atmospheric concentration of carbon dioxide (CO_2), thus reducing greenhouse warming, SRM would either deflect sunlight before it reaches Earth or otherwise cool Earth by increasing the reflectivity of its surface or atmosphere.

In conducting this technology assessment, we focused primarily on the technical status of climate engineering and the views of a wide range of experts on the future of research.[1] Our findings indicate that

- climate engineering technologies are not now an option for addressing global climate change, given our assessment of their maturity, potential effectiveness, cost factors, and potential consequences. Experts told us that gaps in collecting and modeling climate data, identified in government and scientific reports, are likely to limit progress in future climate engineering research.

- the majority of the experts we consulted supported starting significant climate engineering research now. Advocates and opponents of research described concerns about its risks and the possible misuse of its results. Research advocates supported balancing such concerns against the potential for reducing risks from climate change. They further envisioned a future federal research effort that would emphasize risk management, have an international focus, engage the public and national leaders, and anticipate new trends and developments.

- a survey of the public suggests that the public is open to climate engineering research but is concerned about its possible harm and supports reducing CO_2 emissions.

Technical status

To assess the current state of climate engineering technology, we rated each technology for its maturity on a scale of 1 to 9, using technology readiness levels (TRL)—a standard tool for assessing the readiness of emerging technologies before full-fledged production or incorporation into an existing technology or system. We found that climate engineering technologies are currently immature, based on the TRL scores we calculated, and may face challenges with respect to potential effectiveness, cost factors, and potential consequences. (We characterized a technology with a TRL score lower than 6 as immature.)

CDR technologies are designed to do one of the following: (1) chemically scrub CO_2 from the atmosphere by direct air capture, followed by geologic sequestration of the removed CO_2; (2) use biochar and biomass

---

[1] The request for this assessment was originally made during the 111th Congress by the Chairman of the House Committee on Science and Technology, who has since retired.

approaches to capture and sequester CO_2; (3) manage land use to enhance the natural uptake and storage of CO_2; (4) accelerate CO_2 transfer from the atmosphere to the deep ocean for sequestration. We scored all but one CDR technology at a maturity of TRL 2. This means that we found that scientific or government publications have reported

• observation of the technology's basic scientific principles through theoretical research or mathematical models and

• conceptualization of an application of the technology in the context of addressing global climate change—but not an analytic and experimental proof of concept.

The highest-scoring CDR technology (at TRL 3) was direct air capture of CO_2, which has had laboratory demonstrations using a prototype and field demonstrations of underground sequestration of CO_2. However, direct air capture is believed to be decades away from large-scale commercialization. Additionally, for each of the currently proposed CDR technologies, we found that implementation on a scale that could affect global climate change may be impractical, either because vast areas of land would be required or because of inefficient processes, high cost, or unrealistically challenging logistics.

SRM technologies would inject aerosols into Earth's stratosphere to scatter a fraction of incoming sunlight, artificially brighten clouds, place solar radiation scatterers or reflectors in space, or increase the reflectivity of Earth's surface. All SRM technologies' maturity measured TRL 2 or less. That is, none had an analytical and experimental proof of concept. Additionally, we found that the SRM technologies that we rated "potentially fully effective" have not, thus far, been shown to be without possibly serious consequences. Further, each SRM technology must be maintained to sustain its effects on Earth's temperature; discontinuing the technology for any reason would result in Earth's temperature rising to a level dictated by other changes, such as an increased concentration of CO_2 in Earth's atmosphere.

A key challenge in climate engineering research is safely evaluating the technologies' potential risks in advance of large-scale field tests or deployment. Climate modeling would be a helpful evaluative tool, but a number of both federal agency and scientific reports have identified limitations in climate models and their underlying bases. Expanded scientific knowledge, enhanced precision and accuracy of tools for measuring key climate variables, and the development of dedicated high-performance computing would help fill the gaps and make future research more effective.

Future directions

To determine how experts view the future of climate engineering research, we consulted 45 experts with a wide range of backgrounds and professional affiliations. We used future scenarios developed by one set of experts as a foresight tool to help elicit other experts' views. We found that the majority of those we consulted advocated starting significant research now or soon and believed that such research would have the potential to help reduce future risks from climate change. However, some conditioned their advocacy on the continuation of efforts to reduce emissions. Additionally, some pointed to new risks that the research or technologies developed from it might introduce.

Many of the experts we consulted advocated research now because of their anticipation that substantial progress toward effective technologies might require two or more decades. Others said that climate engineering research is needed, even if future climate trends (such as the pace of change) are currently uncertain, because such research represents "an insurance policy against the worst case [climate change] scenarios." Many of those who called for research now saw the situation as urgent, reflecting foresight literature that warns against falling behind a potentially damaging trend—with possibly irreversible and very costly consequences. Their view was that climate engineering research now would constitute timely preparation for action and thus may help minimize the possibility of negative outcomes.

A small number of those we consulted opposed future research on climate engineering. Research opponents reasoned that future climate change will not be great enough to warrant climate engineering or that alternatives such as pursuing emissions reductions (without climate engineering) would be preferable. However, the reason for opposing climate engineering research that was most strongly expressed concerned the risks associated with the research itself or the technologies' deployment.

Both research advocates and opponents cautioned that climate engineering research carries risks either in conducting certain kinds of research or in using the results (for example, deploying potentially risky technologies that were developed on the basis of the research). Some also noted that other nations are conducting research and warned that, in the future, a single nation might unilaterally deploy a technology with transboundary effects. The research advocates suggested managing risks from climate engineering by, for example, conducting interdisciplinary risk assessments, developing norms and best practice guidelines for open and safe research, evaluating deployment risks in advance—and, potentially, as we discuss below, conducting joint research with other countries. Some advocates also indicated that rigorous research could help reduce risks from the uninformed use of risky technologies (as, for example, might occur in a perceived emergency) or emphasized the need to weigh potential risks from climate engineering against risks from climate change.

Research advocates envisioned federal research that would foster developing and evaluating technologies like CDR and SRM and emphasize risk management. The majority of research advocates supported research that would include

- an international focus, sponsoring, for example, joint research with other nations (to foster cooperation and shared norms) and the study of how one nation's deployment might affect others, including those that might respond negatively or be especially vulnerable;

- engagement with the public and U.S. decision-makers that might entail conducting studies to address concerns and support decisions (for example, studies of economic, ethical, legal, and social issues and studies of systemic risks); and

- foresight activities to help anticipate emerging research developments, key trends, and their implications for climate engineering research—notably, the new or emerging opportunities and risks that such changes might bring.

Such features are broadly relevant to risk management in that they might (1) reduce risks of international tensions or even conflict resulting from climate engineering, (2) help prepare the nation in advance of possible crises, and (3) anticipate new risks that might be associated with future technologies.

The United States does not now have a coordinated federal approach to climate engineering research, and we earlier recommended that such an approach be developed in the context of a federal strategy to address climate change (GAO 2010a). Other approaches to addressing climate change include efforts to (1) reduce CO_2 emissions and (2) adapt to climate change.

Potential responses

To understand public opinion, we analyzed survey data from 1,006 adults 18 years old and older selected to represent the U.S. population. We provided them with basic materials on climate engineering—that is, information similar in amount and type to what they might receive in the news media. The materials included a definition and examples of climate engineering technologies. Our survey revealed that a majority of the U.S. population is not familiar with climate engineering but may be open to research.

Once provided with explanatory material, about 50–70 percent of the respondents across a range of demographic groups would be open to research on climate engineering and about 45 percent would be somewhat to extremely optimistic about its benefits. Such optimism would be tempered by caution, as we estimate that about 50–75 percent of the U.S. adult public would be concerned about the technologies' safety. Our survey results also indicate that support for reducing CO_2 emissions is more widespread than support for climate engineering. About 65–75 percent of the public would support the involvement of multiple organizations and interests in decision-making on these technologies. They included the scientific community, a coalition of national governments, individual national governments, the general public, private foundations, and not-for-profit organizations.

Contents

Highlights   i

Letter   iii

Summary   v

1 Introduction   1

2 Background   9

3 The current state of climate engineering science and technology   13

   3.1 Selected CDR technologies   14

      3.1.1 Direct air capture of CO_2 with geologic sequestration   20

      3.1.2 Bioenergy with CO_2 capture and sequestration   23

      3.1.3 Biochar and biomass   25

      3.1.4 Land-use management   26

      3.1.5 Enhanced weathering   27

      3.1.6 Ocean fertilization   28

   3.2 Selected SRM technologies   30

      3.2.1 Stratospheric aerosols   33

      3.2.2 Cloud brightening   35

      3.2.3 Scatterers or reflectors in space   36

      3.2.4 Reflective deserts, flora, and habitats   39

   3.3 Status of knowledge and tools for understanding climate engineering   42

      3.3.1 Better models would help in evaluating climate engineering proposals   42

      3.3.2 Key advancements in scientific knowledge could help improve climate models   44

      3.3.3 Better observational networks could help resolve uncertainties in climate engineering science   44

      3.3.4 High-performance computing resources could help advance climate engineering science   46

4 Experts' views of the future of climate engineering research   49

   4.1 A majority of experts called for research now   50

   4.2 Some experts opposed starting research   53

   4.3 A majority of experts envisioned federal research with specific features   54

   4.4 Some experts thought that uncertain trends might affect future research   58

5 Potential responses to climate engineering research   61

   5.1 Unfamiliarity with geoengineering   62

   5.2 Concern about harm and openness to research   63

   5.3 Views on geoengineering in the context of climate and energy policy   66

   5.4 Support for national and international cooperation on geoengineering   68

6 Conclusions   71

7 Experts' review of a draft of this report   73

    7.1 Our framing of the topic   73

    7.2 Our assessment of the technologies   73

    7.3 Our assessment of knowledge and tools for understanding climate engineering 73

    7.4 Our foresight and survey methodologies   74

8 Appendices   75

    8.1 Objectives, scope, and methodology   75

    8.2 Experts we consulted on climate engineering technologies   89

    8.3 Foresight scenarios   92

    8.4 The six external experts who participated in building the scenarios   99

    8.5 Experts who commented in response to the scenarios   100

    8.6 Experts who participated in our meeting on climate engineering   102

    8.7 Reviewers of the report draft   103

9 References   105

GAO contacts and staff acknowledgments   117

Related GAO products   118

Other GAO technology assessments   118

Figures

    Figure 1.1 The Keeling curve, 1960–2010   1

    Figure 1.2 Earth's carbon cycle   4

    Figure 2.1 Global average energy budget of Earth's atmosphere   9

    Figure 3.1 Capturing and absorbing CO_2 from air   20

    Figure 3.2 CO_2-absorbing synthetic tree   20

    Figure 4.1 Taking early action to avoid potentially damaging trends: Illustration from
                 foresight literature   51

    Figure 5.1 U.S. public support for actions on climate and energy, August 2010   67

    Figure 5.2 U.S. public views on who should decide geoengineering technology's use, August 2010   68

    Figure 8.1 Four scenarios defining alternative possible futures   84

Tables and boxes

    Table 1.1   Selected climate engineering proposals, 1877–1992   6-7

    Table 3.1   Selected CDR technologies: Their maturity and a summary of available information   15-19

    Table 3.2   Selected SRM technologies: Their maturity and a summary of available information   31-32

    Table 5.1   Geoengineering types and examples given to survey respondents   64

    Table 8.1   Nine technology readiness levels described   78-79

    Box 4.1   Climate engineering research: Risk mitigation strategies from the literature   52

Abbreviations

AOGCM	atmosphere-ocean general circulation model
BECS	bioenergy with CO_2 capture and sequestration
CCS	carbon capture and sequestration
CDR	carbon dioxide removal
CLARREO	Climate Absolute Radiance and Refractivity Observatory
DOE	U.S. Department of Energy
EOR	enhanced oil recovery
ESM	Earth system model
GCM	general circulation model
GPU	graphics processing unit
IPCC	Intergovernmental Panel on Climate Change
NAS	National Academy of Sciences
NASA	National Aeronautics and Space Administration
NETL	National Energy Technology Laboratory
NIST	National Institute of Standards and Technology
NOAA	National Oceanic and Atmospheric Administration
NRC	National Research Council
NSF	National Science Foundation
PNNL	Pacific Northwest National Laboratory
R&D	research and development
SRM	solar radiation management
TRL	technology readiness level
USDA	U.S. Department of Agriculture

1 Introduction

Every day, millions of tons of carbon-rich compounds called fossil fuels are extracted, refined or processed, and combusted to supply the world with energy, releasing as a byproduct millions of tons of carbon dioxide gas (CO_2).[2] From 1900 to 2007, annual global CO_2 emissions from fossil fuel consumption increased, on average, at a rate of about 2.6 percent per year (Boden, Marland, and Andres 2010).[3]

As emissions increased, the atmospheric concentration of CO_2 rose. Figure 1.1 shows the rise in the concentration of CO_2 between 1960 and 2010 (Ralph Keeling 2011).[4]

C. D. Keeling (1960), noting that CO_2 levels at observation stations were increasing over time, attributed this increase to fossil fuel combustion.[5] Although CO_2 is not the most abundant

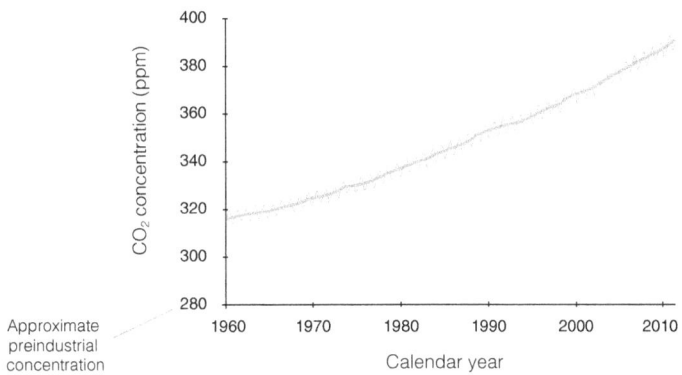

Figure 1.1 The Keeling curve, 1960–2010. Source: GAO, adapted from Ralph Keeling (2011).

The orange line indicates the annual average atmospheric concentration of CO_2 derived from monthly in situ air measurements at Mauna Loa Observatory, Hawaii. The cyclical pattern of the monthly measurements shown in light grey indicates seasonal fluctuations. The approximate preindustrial concentration of 280 parts per million (ppm) indicates the estimated atmospheric abundance of CO_2 around the year 1750. In 1960, the atmospheric concentration of CO_2 was about 317 ppm; by 2010, it had risen to about 390 ppm.

---

[2] Fossil fuels, such as coal, oil, and methane, are natural organic compounds of mostly carbon (C) and hydrogen (H). These fuels are formed from dead plant and animal matter that has been subjected to intense pressure and heat over geologic time scales.

[3] Compound annual growth rate calculated from available emissions estimates.

[4] Over time, atmospheric CO_2 can be reabsorbed as sediment on the ocean floor through the carbon cycle.

[5] In the atmosphere, greenhouse gases absorb and reemit radiation within the thermal infrared range of the electromagnetic spectrum. This is the fundamental cause of the greenhouse effect, or the warming of Earth's atmosphere. In order of their prevalence by volume, the primary greenhouse gases are water vapor (H_2O), CO_2, methane (CH_4), nitrous oxide (N_2O), and ozone (O_3) (Baird 1998).

greenhouse gas, many scientists have concluded that CO_2 emitted by human activities is the principal cause of the enhanced greenhouse effect (Lacis et al. 2010).[6]

Over the past century, global mean surface temperature increased by about 0.75 degrees Celsius, and many scientists expect the rise to continue in coming decades (NRC 2010a; Solomon et al. 2007), as we describe in the background section of this report.[7] A few scientists have argued that a doubling of the atmospheric CO_2 concentration, by itself, would increase the global average temperature by only about 1 degree Celsius and that the models predicting rising temperatures in the coming decades are incomplete and are therefore considerably uncertain (Lindzen 2010; Lindzen and Choi 2009).[8] Nevertheless, there is a consensus of many authoritative scientific bodies, which have conveyed a sense of urgency on the climate change issue; hence the following discussion on climate engineering, or direct,

deliberate large-scale interventions in Earth's climate.[9]

The future effects of warming are uncertain. The National Research Council (NRC) recently examined potential consequences of rising temperatures over the next century, such as changes in vegetation, precipitation, and the rate of sea level rise (NRC 2010a). NRC's report suggests an overall potential for negative effects on people, infrastructures, and ecosystems. For example, the projected rise in sea level could threaten several large ports and urban centers in the United States, such as Miami, New York, and Norfolk, as well as low-lying island groups, such as the Maldives.[10] Some researchers have suggested that climate change could have even more extreme adverse consequences.[11] Others have proposed that rising temperatures might benefit certain geographic areas or economic sectors; for example, agricultural productivity might increase in some areas, although researchers caution that how climate change affects agriculture is complex and uncertain (Gornall et al. 2010). Additionally, while global surface temperature is increasing on average, it is not increasing uniformly (Solomon et al. 2007). For example, scientists have observed that temperatures have risen more in areas that

---

[6] Water vapor (H_2O) is the most abundant greenhouse gas and has a powerful effect on warming (Solomon et al. 2007; Kiehl and Trenberth 1997). Scientists have shown that the tropospheric water vapor concentration significantly affects the global average surface temperature. The enhanced greenhouse effect caused by emissions from human activities is sometimes called anthropogenic climate change. The increased concentration of CO_2 is also known to be the leading cause of another major environmental concern in addition to warming: ocean acidification, manifested by decreases in pH (hydrogen ion concentration), is caused by the oceans' greater uptake of atmospheric CO_2 as its abundance increases (Sabine et al. 2004).

[7] Multiple, interrelated systems can influence the enhanced greenhouse effect.

[8] These scientists argue that all current climate prediction models incorrectly project more warming, based on positive feedback from water vapor and clouds. Specifically, they argue that such feedback has a negative effect (Lindzen and Choi 2009). The Intergovernmental Panel on Climate Change (IPCC) has also noted the uncertainty surrounding such feedback (Solomon et al. 2007).

[9] This report is an assessment of technologies to engineer the climate and the quality of information available to assess these technologies. In this report, we did not assess whether the climate is changing or what is causing any climate change that is occurring or whether current scientific knowledge supports the notion that the climate is changing or its causes. We did not assess whether climate change is or will be sufficient to warrant using these technologies.

[10] Mohamed Nasheed, President of the Maldives, has said that sea level rise is already causing coastal erosion in his country, evidenced by salt intrusion in the water table and relocations affecting 16 islands (Eilperin 2010).

[11] For example, climate change might lead to greater scarcity of food, water, or shelter and social upheaval in many countries in Africa, Asia, and the Middle East (CNA Corporation 2007, 44). Some have suggested that disrupted food and water supplies in certain regions might lead to mass migrations or international conflict (Dyer 2010).

are relatively colder, and the observed change in temperature is greater in winter than in summer and greater at night than in the day (Solomon et al. 2007). Disproportionate warming of cold temperatures could have important implications for human health and mortality, if exposure to heat is less dangerous than exposure to cold (NRC 2010a).

Two broad strategies to meet the challenges of climate change through public policy are mitigation and adaptation. Mitigation aims to limit climate change, usually by decreasing greenhouse gas emissions (GAO 2008a). For example, mitigation might replace high-carbon fuels, such as coal, with fuels that emit less CO_2 per unit of energy, such as natural gas. Mitigation might also enhance the capacity of sinks, which reabsorb CO_2 from the atmosphere and store it on Earth (GAO 2008a). For example, incremental changes in land use could increase the amount of carbon stored as cellulosic fiber in forests and other vegetation that removes CO_2 from the atmosphere by photosynthesis. Adaptation aims to adjust Earth's systems, infrastructures, or social programs in response to actual or expected changes in the climate. For example, adaptation can make systems more robust in the face of climatic extremes, exploit new opportunities, or cope with adversity (GAO 2009a).

Success in mitigating climate change or adapting to it can depend on technological progress. For example, the cost of mitigation is likely to be lower if alternatives to fossil fuels are less expensive (Popp 2006). Adaptation can also be affected by the technology, as happens in predicting the weather, controlling indoor temperatures with heating and air conditioning, or managing a sea level rise, as in building harbor gates in Venice, Italy (Spencer et al. 2005). However, neither mitigation nor adaptation

has progressed sufficiently to moderate current climate projections or diminish the seriousness of their effects. For example, the relative expense of low-carbon energy technology presently tends to limit its use. And requirements to reduce emissions can be difficult to enforce, as the Kyoto Protocol demonstrates, or can fail to encourage advances in low-carbon energy technology (Barrett 2008; Barrett 1998).

Even if deep emissions cuts were to stabilize the atmospheric concentration of CO_2 at the current level, scientific models predict that average global surface temperature is likely to rise 0.3 to 0.9 degrees Celsius by 2100 (Backlund et al. 2008). Some scientists suggest that climatic perturbation from anthropogenic CO_2 emissions is nearing a tipping point beyond which it will be difficult or impossible to remediate changes in Earth's climate. Figure 1.2 illustrates Earth's carbon cycle, which regulates the flow of carbon between the atmosphere and land-based and oceanic sinks.

These and other possible challenges to the success of mitigation and adaptation have helped stimulate public policy interest in climate engineering, which would develop and use technology to moderate Earth's climate by controlling the radiation balance and, thus, average global temperature. The United Kingdom's Royal Society (the oldest scientific academy in continuous existence) has identified other distinguishing characteristics of this strategy as well, highlighting the "deliberate, large-scale intervention in the Earth's climate system" in its definition of geoengineering (Royal Society 2009, ix).[12]

---

[12] We use the term "geoengineering" in appropriate contexts, as when it refers to information we collected in a survey of U.S. adults and their attitudes toward technologies to address climate change. We described the alternative terms "climate engineering," "climate remediation," and "climate intervention" in a September 2010 report (GAO 2010a, 3).

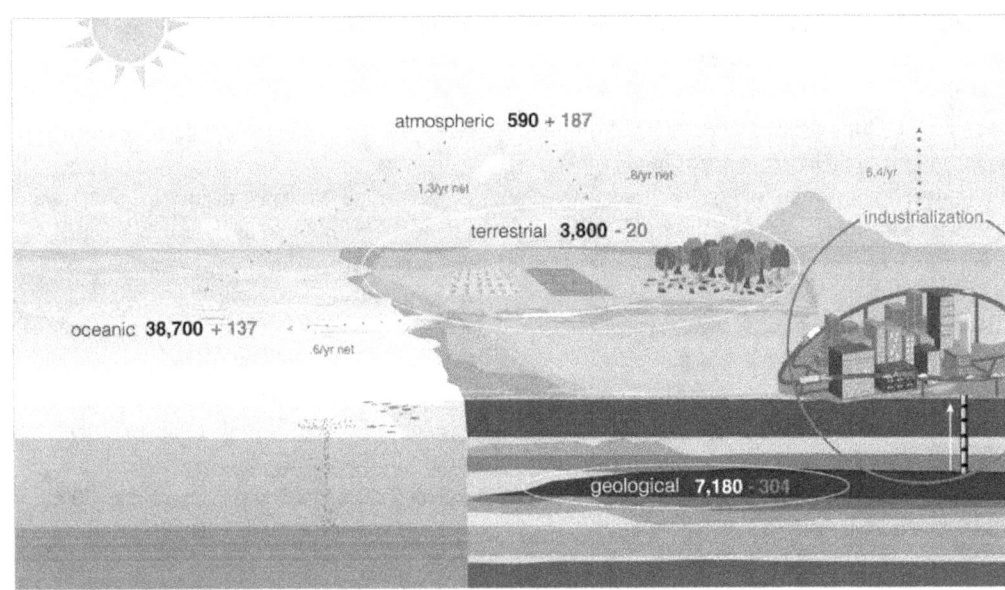

Figure 1.2 Earth's carbon cycle. Source: GAO, adapted from Sarmiento and Gruber (2002), updated using Field, Sarmiento, and Hales (2007).

Note: All numeric values are in gigatons (GtC), or billions of metric tons, of carbon. In Earth's carbon cycle, preindustrial reservoir sizes are represented by black numbers. Cumulative postindustrial reservoir transfers are represented by red numbers. Current fluxes between reservoirs are shown in smaller type; the largest flux is 6.4 GtC per year from industrialization. This ongoing carbon imbalance is causing ocean water to become more acidic and is believed to be the primary cause of increased global average surface temperature. (An animated depiction of changes in the global carbon cycle over time may be accessed at www.gao.gov/multimedia/interactive/GAO-11-71a.)

In its 2009 report, the Royal Society described two major approaches to climate engineering: accelerating the movement of carbon from the atmosphere to terrestrial and oceanic carbon sinks, or carbon dioxide removal (CDR), and controlling net incoming radiation from the Sun, or solar radiation management (SRM). As CDR reduces the atmospheric concentration of CO_2, the enhanced greenhouse effect is weakened, and thermal radiation more easily escapes into space.[13] SRM, in contrast, attempts to reduce net incoming solar radiation by deflecting sunlight or by increasing the reflectivity of the atmosphere, clouds, or Earth's surface.[14]

The concept of engineering the climate is not new (Fleming 2010). Table 1.1 shows examples of climate engineering proposals dating from 1877. Today, policymakers and scientists are examining climate engineering as a way to manage potential catastrophic risks from climate change.

[13] Although experts differ on which technologies to define as climate engineering (Gordon 2010, ii), in this report we limited our assessment to key climate engineering technologies among those reviewed by the Royal Society (Royal Society 2009).

[14] Because SRM would not affect the atmospheric concentration of CO_2, it would not abate increased ocean acidification.

We designed this report to complement our September 2010 report on geoengineering (GAO 2010a). In this context, we conducted this technology assessment of climate engineering.[15] Our objectives for this report were to examine (1) the current state of climate engineering science and technology, (2) expert views of the future of U.S. climate engineering research, and (3) public perceptions of climate engineering (we describe our methodology in section 8.1).

To determine the current state of the science and technology of climate engineering, we reviewed a broad range of scientific and engineering literature, including proceedings from conferences such as the 2010 Asilomar International Conference on Climate Intervention Technologies (Asilomar Scientific Organizing Committee 2010). We revisited GAO-10-903, a complementary report on climate engineering we issued in September 2010 (GAO 2010a). We reviewed relevant congressional testimony. We interviewed a broad range of experts and officials working on climate engineering and proponents of specific climate engineering technologies. This report is an

assessment of technologies to engineer the climate and the quality of information available to assess these technologies. We did not independently assess whether climate change is occurring or what is causing any climate change if it is occurring or whether current scientific knowledge supports the occurrence of climate change or its causes. We did not assess whether climate changes are or will be sufficient to warrant using these technologies.

To ensure a balance of views and information, we analyzed and synthesized information from an array of experts with diverse views on our subject. We used the Royal Society's classification of climate engineering approaches to focus our analysis on CDR and SRM technologies (Royal Society 2009, 1). From the information we found in the literature and our interviews with experts, we assessed climate engineering technologies along four key dimensions: (1) maturity, (2) potential effectiveness, (3) cost factors, and (4) potential consequences. We did not independently assess the accuracy of the cost estimates, but we report estimates we found in the literature.

To assess how experts view the future of climate engineering research, we (1) conducted a foresight exercise in which experts developed alternative future scenarios; (2) elicited comments, stimulated by the scenarios, from a broad array of experts; and (3) asked other experts to respond to the preliminary synthesis we developed from the scenarios and earlier comments.

---

[15] In the Senate report accompanying the proposed bill for the legislative branch fiscal year 2008 appropriation, the Senate Committee on Appropriations recommended the establishment of a permanent technology assessment function within GAO (United States Senate 2007, see S. Rep. No. 110-89, at 42–43 (2007)). The House Committee on Appropriations, in providing funding to GAO to perform technology assessment studies, noted that "it is necessary for the Congress to equip itself with effective means for securing competent, timely and unbiased information concerning the effects of scientific and technical developments and use the information in the legislative assessment of matters pending before the Congress" (U.S. House of Representatives 2007, see H.R. Rep. No. 110-198, at 30 (2007)). GAO established a permanent operational technology assessment group within its Applied Research and Methods team: the Center for Science, Technology, and Engineering. GAO defines technology assessment as the thorough and balanced analysis of significant primary, secondary, indirect, and delayed interactions of a technological innovation with society, the environment, and the economy and the present and foreseen consequences and effects of those interactions.

Date	Who	Proposal
1877	Nathaniel Shaler, American scientist	Suggested rerouting the Pacific's warm Kuroshio Current through the Bering Strait to raise Arctic temperatures as much as 30 degrees Fahrenheit
1912	Carroll Livingston Riker, American engineer, and William M. Calder, U.S. Senator	Proposed building a 200-mile jetty into the Atlantic Ocean to divert the warm Gulf Stream over the colder Labrador current to change the climate of North America's Atlantic Coast; Calder introduced a bill to study its feasibility
1929	Hermann Oberth, German-Hungarian physicist and engineer	Proposed building giant mirrors on a space station to focus the Sun's radiation on Earth's surface, making the far North habitable and freeing sea lanes to Siberian harbors
1945	Julian Huxley, biologist and Secretary-General of UNESCO 1946–48	Proposed exploding atomic bombs at an appropriate height above the polar regions to raise the temperature of the Arctic Ocean and warm the entire climate of the northern temperate zones
c. 1958	Arkady Markin, Soviet engineer	Proposed that the United States and Soviet Union build a gigantic dam across the Bering Strait and use nuclear power–driven propeller pumps to push the warm Pacific current into the Atlantic by way of the Arctic Sea. Arctic ice would melt, and the Siberian and North American frozen areas would become temperate and productive
1958	M. Gorodsky, Soviet engineer and mathematician, and Valentin Cherenkov, Soviet meteorologist	Proposed placing a ring of metallic potassium particles into Earth's polar orbit to diffuse light reaching Earth and increase solar radiation to thaw the permanently frozen soil of Russia, Canada, and Alaska and melt polar ice
1965	President's Science Advisory Committee, United States	Investigated injecting condensation or freezing nuclei into the atmosphere to counteract the effects of increasing carbon dioxide
1977	Cesare Marchetti, Italian industrial physicist	Coined the term "geoengineering" and proposed sequestering CO_2 in the deep ocean

Table 1.1 Selected climate engineering proposals, continues on next page

Date	Who	Proposal
1983	Stanford Penner, A. M. Schneider, and E. M. Kennedy, American physicists	Suggested introducing small particles into the atmosphere to reflect more sunlight back into space
1988	John H. Martin, American oceanographer	Proposed dispersing a relatively small amount of iron into appropriate areas of the ocean to create large algae blooms that could take in enough atmospheric carbon to reverse the greenhouse effect and cool Earth
1989	James T. Early, American climatologist	Suggested deflecting sunlight by 2 percent with a $1 trillion to $10 trillion "space shade" placed in Earth orbit
1990	John Latham, British cloud physicist	Proposed seeding marine stratocumulus clouds with seawater droplets to increase their reflectivity and longevity
1992	NAS Committee on Science, Engineering, and Public Policy	Proposed adding more dust to naturally occurring stratospheric dust to increase the net reflection of sunlight

Table 1.1 Selected climate engineering proposals, 1877–1992. Source: GAO.

Note: Table 1.1 is based in part on an outline provided by James R. Fleming. We selected proposals beginning in the 19th century to illustrate a variety of climate engineering technologies and points in Earth's climate system where interventions could occur. The table excludes numerous proposals to generate rain or alter hurricanes, which are not intended to cause long-term change.

We also conducted focus groups and a web-based survey of the U.S. adult population. We surveyed a representative sample of U.S. residents 18 years old and older from July 19 to August 5, 2010, receiving usable responses from 1,006 respondents. We used the term "geoengineering" in the information we gave the focus group and survey participants, given that we and others, such as the Royal Society, had used this term earlier.

We convened a meeting of scientists, engineers, and other experts, with the assistance of the National Academy of Sciences (NAS), that we called the Meeting on Climate Engineering.

We helped NAS select a diverse and balanced group of participants with expertise in climate engineering, climate science, measurement science, foresight studies, emerging technologies, research strategies, and the international, public opinion, and public engagement dimensions of climate engineering. We provided them with the preliminary results of our work, and the meeting served as a forum in which the participants expressed general reactions to and gave advice and suggestions on our preliminary findings. Their comments led us to review additional published and unpublished literature.

Following the meeting, we contacted the participants in person or by telephone or e-mail to clarify and expand what we had heard. We used what we learned from the meeting participants to update and clarify our exposition of the current state of climate engineering technology, expert views of the future of U.S. climate engineering research, and public perceptions of climate engineering. We then sent a complete draft of our report to the participants in the Meeting on Climate Engineering who had agreed to review it.

We conducted our work for this technology assessment from January 2010 through July 2011 in accordance with GAO's quality standards as they pertain to technology assessments. Those standards require that we plan and perform the technology assessment to obtain sufficient and appropriate evidence to provide a reasonable basis for our findings and conclusions, based on our technology assessment objectives. We believe that the evidence we obtained provides a reasonable basis for our findings and conclusions, based on our technology assessment objectives.

2 Background

Global temperature increases such as those measured on Earth have been attributed to a gradual change in the balance of energy flowing into and away from Earth's surface. Earth's system maintains a constant average temperature only if the same amount of energy leaves the system as enters it. If more energy enters than leaves, the difference manifests as a temperature increase. Figure 2.1 shows current estimates of the equilibrium transfer of energy.

Solar radiation is the predominant source of energy entering Earth's system. It has an average global power of approximately 342 watts per square meter (W/m²). The system, including Earth's surface and the atmosphere, absorbs about 69 percent of incoming solar radiation and reflects the remaining 31 percent back into space. That is, Earth's surface absorbs about 49 percent of incoming radiation, and the atmosphere absorbs about 20 percent. Earth's atmosphere

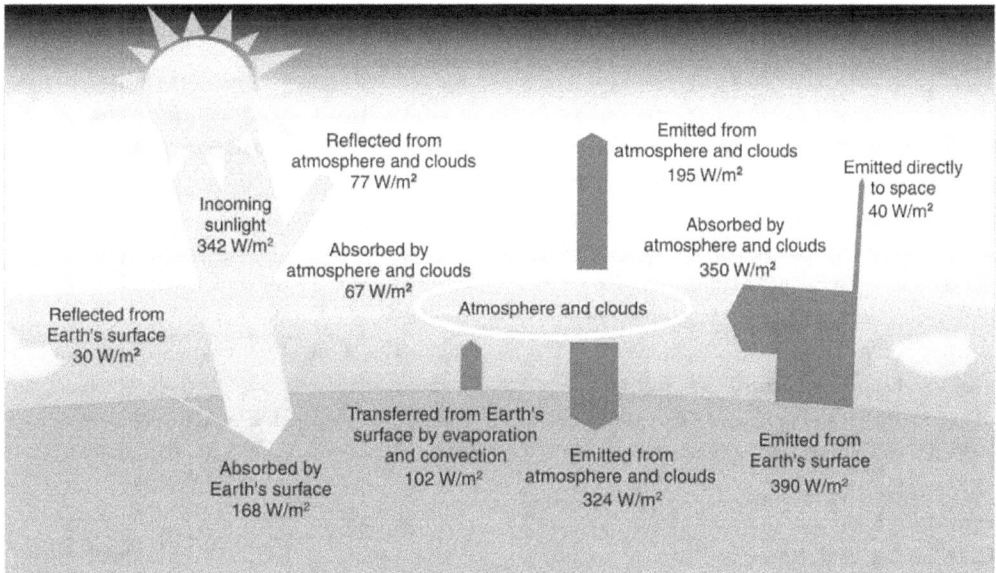

Figure 2.1 Global average energy budget of Earth's atmosphere. Source: GAO, adapted from Kiehl and Trenberth 1997.

Note: All numeric values are in watts per square meter (W/m²). Incoming sunlight is both reflected from and absorbed by the atmosphere, clouds, and Earth's surface. Some of the energy absorbed by Earth's surface is transferred to the atmosphere by evaporation and convection, and the remainder is emitted as heat energy. The majority of the heat energy is absorbed by the atmosphere and clouds, with some escaping directly to space. Energy absorbed by the atmosphere and clouds is reradiated as heat energy back to Earth's surface as well as directly to space. Based on the composition of the atmosphere and clouds, the heat energy they absorb can accumulate by the greenhouse effect in which energy emitted from Earth's surface is trapped by gases in the atmosphere and clouds. For this reason, greenhouse gases in Earth's atmosphere can affect global average surface temperature. (An animated depiction of the global average energy budget of Earth's atmosphere may be viewed at www.gao.gov/multimedia/interactive/GAO-11-71b.)

and clouds reflect approximately 23 percent into space, while Earth's surface (land, vegetation, water, and ice) reflects approximately 9 percent. Energy absorbed by the atmosphere affects the planet's climate system through subsequent energy transfers (Solomon et al. 2007).

The energy Earth's surface and atmosphere absorb warms the planet. An inflow of energy to Earth without an equivalent outflow would result in continually increasing temperatures. However, Earth reemits energy from the surface to the atmosphere in the form of thermal radiation (long wavelength or infrared radiation) (Solomon et al. 2007).

Approximately 10 percent of the thermal radiation reemitted by Earth passes through the atmosphere into space, and 90 percent is absorbed in the atmosphere, primarily in greenhouse gases, which efficiently absorb long-wave radiation. The atmospheric concentration of greenhouse gases is very low. Water vapor (H_2O) is the most important greenhouse gas and is highly variable but typically makes up about 1 percent of the atmosphere (Solomon et al. 2010; Kiehl and Trenberth 1997). Carbon dioxide is the second most important greenhouse gas; the current atmospheric concentration of CO_2 is approximately 390 ppm (R. F. Keeling et al. 2009; Kiehl and Trenberth 1997; C. D. Keeling et al. 2001).[16]

Just as the planet must maintain a balance of incoming and outgoing energy, the atmosphere and clouds must emit as much energy as they absorb to maintain a constant temperature. Therefore, the atmosphere and clouds emit

long-wave radiation at approximately the same rate as they absorb energy from the Sun and Earth. This is manifested as additional thermal emissions both into space and toward Earth. The planet's surface absorbs the Earth-bound thermal radiation, which raises Earth's surface temperature, which increases thermal radiation from Earth's surface, and so on, until this feedback achieves stable temperatures.

The relationship between temperature and thermal radiation emitted from Earth is approximately described by the Stefan-Boltzmann law:

$$F = \sigma T^4$$

where F is the thermal radiation emitted from Earth's surface in watts per square meter (W/m^2), σ is the Stefan-Boltzmann constant, and T is the temperature of Earth's surface in Kelvin (K).[17] The Stefan-Boltzmann law provides evidence for atmospheric greenhouse gas feedback in Earth's energy system. If Earth's radiation, absorbed and reemitted, were only 235 W/m^2 (342 W/m^2 minus 107 W/m^2 of reflected solar radiation), its average surface temperature would be about 254 K (−19 degrees Celsius). But Hansen and colleagues have estimated that Earth's actual average surface air temperature between 1951 and 1980 was approximately 287 K (14 degrees Celsius) (Hansen et al. 2010). The difference in temperature is attributed to greenhouse gases that trap thermal radiation, warming Earth as depicted in figure 2.1. Thermal radiation emitted

---

[16] Energy is also transferred mechanically (not by radiation) from Earth's surface to the atmosphere and clouds by evaporation and convection.

[17] The Stefan-Boltzmann law, named after Jožef Stefan and Ludwig Boltzmann, states that the total power radiated per unit of surface area of a black body per unit of time is directly proportional to the fourth power of the black body's thermodynamic temperature T. The Stefan–Boltzmann constant σ is equal to 5.6704 x 10^{-8} watts per square meter per absolute temperature measured in Kelvin to the fourth power ($W/m^2/K^4$).

by Earth's surface at 287 K is 385 W/m^2, which compares favorably with the 390 W/m^2 in the figure, corresponding to a temperature of 288 K.

Climate scientists infer that accumulations of anthropogenic greenhouse gases are gradually adding to Earth's natural greenhouse process. These accumulations absorb more thermal radiation emitted by Earth's surface and reduce thermal radiation that escapes into space. The additional thermal radiation the greenhouse gases absorb is reradiated to space and back toward Earth. The planet's surface absorbs the additional Earth-bound thermal radiation, which raises Earth's surface temperature, which increases thermal radiation from Earth's surface, and so on, until this feedback achieves a new, higher stable temperature. The magnitude and effect of this change in Earth's global energy system are important subjects of climate science studies today (Solomon et al. 2007; NRC 2010a).

3 The current state of climate engineering science and technology

Most climate engineering proposals would aim to remediate the climate by affecting Earth's energy balance, using either CDR to reduce the atmospheric concentration of CO_2 or SRM to reduce incoming solar radiation. These two approaches differ significantly in their technical challenges and potential consequences (Royal Society 2009). The literature and our interviews with experts suggested four key dimensions on which we assessed these technologies, to the extent possible, given their current development: (1) maturity, (2) potential effectiveness, (3) cost factors, and (4) potential consequences (see section 8.1). Since developing many of the technologies we examined would require advances in new scientific data and analyses, we identified the climate's representative physical, chemical, and biological algorithms; the geographic, temporal, and technical sensors of essential climate mechanisms; and next-generation, high-performance computing resources dedicated to climate science as areas that represent current shortfalls in knowledge and infrastructure.

CDR technologies may be characterized as predominantly land-based or predominantly ocean-based (NRC 2010a; Royal Society 2009). Land-based technologies include direct air capture, bioenergy with CO_2 capture and sequestration, biochar and other biomass-related methods, land-use management, and enhanced weathering. Direct air-capture systems attempt to capture CO_2 from air directly and then store it in deep subsurface geologic formations. Bioenergy with CO_2 capture and sequestration would also store CO_2 underground, and biochar and other biomass-related methods would sequester carbon in soil or bury it. Land-use management practices we reviewed would enhance natural sequestration of CO_2 in forests. Enhanced weathering would fix atmospheric CO_2 in silicate rocks in a chemical reaction and then store it as either carbonate rock or dissolved bicarbonate in the ocean. Ocean-based technologies would fertilize the ocean to promote the growth of phytoplankton to sequester CO_2.

Seven SRM technologies have been reported in sufficient detail for us to assess them as candidates for climate engineering. Two would be deployed in the atmosphere—one scattering solar radiation back into space using stratospheric aerosols, the other reflecting solar radiation by brightening marine clouds. Two would be deployed in space—one scattering or reflecting solar radiation from Earth orbit, the other scattering or reflecting solar radiation at a stable position between Earth and the Sun. The three remaining technologies would artificially reflect additional solar radiation from Earth's surfaces—covered deserts, more reflective flora, or more reflective settled areas.

We found that since most climate engineering technologies are in early stages of development, none could be used to engineer the climate on a large scale at this time. We used technology readiness levels to rate the maturity of each technology on a scale from 1 to 9, with scores lower than TRL 6 indicating an immature technology. No CDR technology scored higher than TRL 3, and no SRM technology scored higher than TRL 2.[18]

Considerable uncertainty surrounds the potential effectiveness of the technologies we reviewed, in part because they are immature. Additionally, for several proposed CDR technologies, the amount of CO_2 removed may be difficult to verify through modeling or direct measurements.

The technologies' cost factors we report represent, for CDR, resources used to remove CO_2 from the atmosphere and store it. For SRM, they represent resources required to counteract global warming caused by doubling the preindustrial atmospheric concentration of CO_2 or, for technologies that are potentially not fully effective, resources required to counteract global warming to the maximum extent possible. Some of the studies we reviewed indicate possible cost levels; we report these for illustration, but we did not evaluate them independently. Some studies described cost levels qualitatively (Royal Society 2009).

Using many of the CDR and SRM technologies we reviewed would pose risks, some of which might not yet be known. Although minimal risks have been reported for air capture, some risks are related to the geologic sequestration of CO_2. Land-use management approaches to capture and store CO_2 are not generally regarded as risky. Enhanced weathering would pose environmental risks from the large-scale mining activities that would be needed to support it. The short-term and long-term ecological, economic, and climatologic risks from ocean fertilization remain uncertain. Using SRM technologies could affect temperatures but would not abate ocean acidification. Potential effects on precipitation are varied. Failing to sustain SRM technologies, once deployed, could result in a potentially rapid temperature rise.

In sections 3.1 and 3.2, we present our assessment of the CDR and SRM technologies. In section 3.3, we describe the status of scientific knowledge and infrastructure related to climate engineering technologies.

3.1 Selected CDR technologies

Table 3.1 summarizes our assessment of the maturity of six CDR technologies and presents information from published reports on their potential effectiveness, cost factors, and potential consequences. TRL ratings assess the maturity of each technology. Potential effectiveness is described in terms of an overall qualitative rating, where possible, and quantitative estimates of (1) the maximum capacity to reduce the global atmospheric concentration of CO_2 (ppm) from its projected level of 500 ppm in 2100 and (2) the annual capacity to remove CO_2 from Earth's atmosphere (gigatons of CO_2 or CO_2-C equivalent per year), which we compared to annual anthropogenic emissions of 33 gigatons of CO_2.[19] Cost factors represent the resources used to remove CO_2 from the atmosphere and store it. Potential consequences associated with each technology include reported negative consequences, risks, and cobenefits.

---

[18] We used the *AFRL Technology Readiness Level Calculator* to assess maturity (see section 8.1). For a rating of TRL 2 or higher, the basic requirement is a system concept on a global scale; for a rating of TRL 3 or higher, analytical and experimental demonstration of proof of concept is required, and for a rating of TRL 4 or higher, system demonstration with a breadboard unit is required. These requirements apply regardless of a technology's scientific basis or the extent to which the techniques it incorporates are well established.

[19] In 2010, the atmospheric concentration of CO_2 was about 390 ppm; around the year 1750, it was about 280 ppm.

Table 3.1 Selected CDR technologies, continues on next page

Technology	Maturity[a]	Potential effectiveness[b]	Cost factors[c]	Potential consequences[d]
Direct air capture of CO_2 with geologic sequestration	**Low (TRL 3):** • Basic principles understood and reported • System concept formulated • Experimental proof of concept demonstrated with a prototype unit in a laboratory environment • Models of CO_2 injection and transport developed and used for risk analysis and for simulating fate of injected CO_2 • Basic technological components not demonstrated as working together • No plans or prototypes for large-scale industrial implementation • Geological sequestration of CO_2 is more mature but not practiced on a scale to potentially affect climate	**Not rated:** • No "obvious limit" to the amount of CO_2 reduction by year 2100 • Could theoretically counter all global anthropogenic CO_2 emissions at 33 gigatons per year • Large energy penalty: net increase in CO_2 emissions if fossil fuel used (electricity from fossil fuels would release more CO_2 than an air capture unit would remove) • Uncertainty around technical scalability	• Viability may depend on nature and extent of a carbon market • Process energy requirements for currently inefficient technologies for directly separating CO_2 from air in very dilute concentration • Transportation and logistics for sequestration of captured CO_2 • Construction and management of geologic CO_2 sequestration sites (e.g., CO_2 injection, measuring, monitoring, and verification) • Greatly varied estimates in the scientific literature: $27 to $630 or more per ton of CO_2 removed (excluding transportation, sequestration, and other costs)	• Aspects associated with handling process materials or chemicals • May have sequestration risks such as potential for CO_2 to escape from underground storage in the event of reservoir fracture or fissure from built-up pressure

Technology	Maturity[a]	Potential effectiveness[b]	Cost factors[c]	Potential consequences[d]
Bioenergy with CO$_2$ capture and sequestration	**Low (TRL 2):** • Basic principles understood and reported • System concept formulated • No experimental demonstration of proof of concept (no laboratory scale experiments that indicate CO$_2$ reducing potential) • Emerging technology leverages what is known about CO$_2$ capture and geologic sequestration	**Low to medium:** • Maximum ability to reduce atmospheric CO$_2$: 50–150 ppm by 2100 • Net carbon negative under ideal conditions • Depends on plant productivity and land area cultivated	• Viability may depend on nature and extent of a carbon market • Value of land in other uses • Potentially large land area for growing and harvesting biomass • Type of biomass feedstock (e.g., switchgrass) • Process energy requirements for bioenergy production (e.g., pyrolysis) • Construction and management of geologic CO$_2$ sequestration sites (e.g., CO$_2$ injection, measuring, monitoring, and verification) • Transportation and logistics for sequestering captured CO$_2$ • Greatly varied estimates in the scientific literature: $150–$500 per ton of CO$_2$ removed (excluding transportation and sequestration costs)	• Potential land-use trade-offs; related impacts on food prices, water resources, fertilizer use • CO$_2$ sequestration risks same as direct air capture

Table 3.1 Selected CDR technologies, continues on next page

Technology	Maturity[a]	Potential effectiveness[b]	Cost factors[c]	Potential consequences[d]
Biochar and biomass methods	**Low (TRL 2):** • Basic principles understood and reported • System concept formulated • Proof of concept shown in modeling and experimental results demonstrating its CO_2 capturing ability–but CO_2 sequestration aspects uncertain • Not practiced on a scale to affect climate. No plans or prototypes for large-scale implementation • Substantial uncertainties about capacity to reduce net emissions of CO_2	**Low:** • Maximum ability to reduce atmospheric CO_2; 10–50 ppm by 2100 • Maximum annual sustainable reduction: 1–2 gigatons CO_2-C equivalent of CO_2, CH_4, and N_2O • Net carbon negative under ideal conditions (comparable to bioenergy with CO_2 capture and sequestration)	• Viability may depend on nature and extent of a carbon market • Soil fertility outcomes • Type of pyrolysis feedstock and related factors • Process energy requirements for bioenergy production (e.g., pyrolysis) • Greatly varied estimates in the scientific literature: $2–$62 per ton of CO_2 removed	• Potential land-use trade-offs • Long-term effects on soil uncertain • Health and safety of pyrolysis and biochar handling • Local benefits to soil enhance crop yield

Table 3.1 Selected CDR technologies, continues on next page

Technology	Maturity[a]	Potential effectiveness[b]	Cost factors[c]	Potential consequences[d]
Land-use management (reforestation, afforestation, or reductions in deforestation)	**Low (TRL 2):** • Basic principles understood and reported • Techniques well established • System concept formulated and estimates of its carbon mitigation potential reported based on modeling studies • No experimental demonstration or proof of systemwide concept of CO_2 capture and sequestration by land-use activities • Not practiced on a scale to affect climate. No plans for large-scale implementation	**Low to medium:** • Potential removal of 1.3–13.8 gigatons CO_2 annually • 0.4–14.2 metric tons of CO_2 sequestered per acre per year • Possible rerelease of sequestered CO_2	• Viability may depend on nature and extent of a carbon market • Value of land in other uses • Potentially large land area for growing or preserving forests • Type of flora planted or preserved • Natural resource requirements for maintenance and management of forests (e.g., water) • Measuring, monitoring, and verification	• Potential land-use trade-offs • Possible cobenefits such as reduced water runoff
Enhanced weathering	**Low (TRL 2):** • Basic principles understood and reported • System concept formulated • No experimental demonstration of proof of system-wide concept • Not practiced on a scale to affect climate. No plans or prototypes for large-scale implementation	**Not rated:** • Limited studies in literature • Some estimates based on models but varied conclusions about levels of effectiveness	• Viability may depend on nature and extent of a carbon market • Design and implementation of silicate-based weathering scheme, including distribution and delivery of material • Mining and transportation of silicate rock, and logistics • Greatly varied estimates in the scientific literature: $4–$100 per ton of CO_2 removed	• Potentially undesirable environmental and other consequences from large-scale mining and transportation

Table 3.1 Selected CDR technologies, continues on next page

Technology	Maturity[a]	Potential effectiveness[b]	Cost factors[c]	Potential consequences[d]
Ocean fertilization	**Low (TRL 2):** • Basic principles understood and reported • System concept formulated • Limited small-scale field experiments conducted but results unclear • Published research mainly theoretical • Not practiced on a scale to affect climate. No plans or prototypes for large-scale implementation	**Low:** • Maximum ability to reduce atmospheric CO_2; 10–30 ppm by 2100 • Scientific uncertainty surrounding (1) duration of carbon sequestered in the ocean, (2) how ecological impacts might limit effectiveness, and (3) how often iron would need to be added • Outcomes from limited experiments not understood or well documented	• Viability may depend on nature and extent of a carbon market • Design and implementation of ocean fertilization scheme, including distribution and delivery of material • Mining and transportation of iron ore, and logistics • Greatly varied estimates in the scientific literature: $8–$80 per ton of CO_2 removed	• Ecological effect on ocean not well understood • Risk of algal blooms causing anoxic zones in the ocean

Table 3.1 Selected CDR technologies: Their maturity and a summary of available information. Source: GAO.

[a] In this report, we considered each technology's maturity in terms of its readiness for application in a system designed to address global climate change. To do this, we used technology readiness levels (TRL), a standard tool that some federal agencies use to assess the maturity of emerging technologies. We characterized technologies with TRL scores lower than 6 as "immature" (section 8.1). The TRL rating methodology considers the maturity level of the whole integrated system rather than individual components of a particular technology.

[b] We assessed potential effectiveness by considering the qualitative judgments of the Royal Society and reported estimates of two quantitative measures: (1) maximum ability to reduce the atmospheric CO_2 (ppm) projected for 2100 and (2) annual capacity to remove CO_2 from Earth's atmosphere (gigatons of CO_2 or CO_2-C equivalent per year). Additionally, we reviewed scientific literature with respect to these measures of effectiveness and for assessments indicating the feasibility of implementing CDR technologies on a global scale to achieve a net reduction of atmospheric CO_2 concentration. A technology was not assigned an overall qualitative rating when there were substantial uncertainties in the literature about its effectiveness (see section 8.1).

[c] Cost factors are resources a system uses to remove CO_2 from the atmosphere and store it. Some of the studies we reviewed indicated possible cost levels, which we provide here for illustration. We did not evaluate this information independently.

[d] Includes potential consequences, risks, and cobenefits.

3.1.1 Direct air capture of CO_2 with geologic sequestration

3.1.1.1 What it is

Direct air capture would chemically scrub CO_2 directly from the atmosphere. In some conceptual designs, air is brought into contact with a CO_2-absorbing liquid solution containing sodium hydroxide or with a solid sorbent in the form of a synthetic ion-exchange resin that selectively absorbs CO_2 gas.[20] Figures 3.1 and 3.2 illustrate two different air-capture units. Figure 3.1 shows an artist's rendering of the air-contactor design, and figure 3.2 illustrates a CO_2-absorbing synthetic tree made from a proprietary resin. A CO_2-absorbing resin (sorbent material) could be shaped as a tree or as packing material placed inside a large column where it would be brought into contact with air. The CO_2-rich solution or synthetic resin would be sent to a regenerator, where the CO_2 would be separated from the liquid by thermal cycling or by exposure to humid air. The resulting concentrated stream of CO_2 could be compressed to liquid form and delivered (by trucks, ships, or pipelines) to a sequestration site.[21] The sorbent would be recycled to capture additional CO_2.

Experts have proposed the compression and transportation of captured CO_2 for sequestration in deep underground geologic or saline formations. Most candidate geologic formations consist of layers of porous underground rock capped by layers of nonporous rock that would keep the injected fluids trapped in the lower pore spaces. The CO_2 would be compressed under elevated pressure (greater than 2,000 psi,

or 13 megapascals (MPa)) and sequestered at the capture site, on shore, or in the deep ocean, where the hydrostatic head of the sea water above would keep the CO_2 from rising to the surface (DOE 2006).[22]

Note: This is a virtual rendering of an air-capture unit designed by Carbon Engineering Ltd. Each such unit would capture about 100,000 tons of CO_2 per year. A battery of such units is intended to work with a chemical recovery plant to produce high-purity CO_2.

Figure 3.1 Capturing and absorbing CO_2 from air.
Source: Carbon Engineering Ltd.

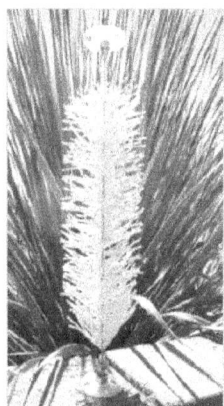

Note: This is a synthetic tree made from a proprietary resin that can absorb CO_2 from air.

Figure 3.2 CO_2-absorbing synthetic tree.
Source: Columbia University.

---

[20] Sorbent refers to a solution or solid that selectively absorbs a specific gas.

[21] Geologic sequestration of CO_2 is a relatively new idea.

[22] Psi indicates pounds per square inch. A megapascal is 1 million pascals; a pascal is a measure of force per unit area, defined as 1 newton per square meter. A newton is the force that produces an acceleration of 1 meter per second per second when exerted on a mass of 1 kilogram. Atmospheric pressure at sea level is 14.7 psi, or roughly 0.1 MPa.

3.1.1.2 Maturity and potential effectiveness

We assessed the maturity of direct air capture of CO_2 with geologic sequestration at TRL 3, given that the basic principles have been observed and reported, a system concept has been formulated, and the literature shows proof of concept—that is, the technology has had laboratory demonstrations using a prototype unit. Direct air capture of CO_2 is probably decades away from commercialization, even though its fundamental chemistry and processes are well understood and laboratory-scale direct air-capture demonstrations are supported at two universities. According to the literature, direct air capture could theoretically remove total annual global anthropogenic CO_2 emissions, estimated at approximately 33 gigatons. The Royal Society reported that this technology had no "obvious limit" to the amount of CO_2 it could capture from the atmosphere. Large-scale implementation, however, is currently neither cost-effective nor thermodynamically efficient. The main difficulty with direct air capture is in the removal of atmospheric CO_2 in its extremely low concentration (approximately 390 ppm), which lowers the thermodynamic efficiency of the process (Ranjan 2010).[23] This would make air capture even more challenging than, for example, capturing CO_2 from a flue stack where the thermodynamic efficiencies were comparatively much higher (approximately 20 percent), mainly because of the higher concentration of CO_2 in the flue gas (about 12 percent or approximately 120,000 ppm).

The low atmospheric CO_2 concentration presents other difficulties such as a significantly large energy penalty associated with the CO_2 absorption system for air capture (Herzog 2003). The total energy required to capture a unit of CO_2 from air is such that if carbon-based fuels such as coal were used as the energy source, more CO_2 would be released to the environment than removed (Zeman 2007). The energy process requirements for the direct air capture of CO_2 would thus have to come from noncarbon or low carbon energy sources. Hence, substantial uncertainties surround the scalability of air capture.

Our interviews with National Energy Technology Laboratory (NETL) engineers revealed that the capacity for sequestering CO_2 in deep underground saline formations is vast enough to store essentially all CO_2 emissions from coal-fired power plants within the United States.[24] Carbon dioxide injection in subsurface geologic formations has been used for decades in enhanced oil recovery (EOR) to extract additional oil from depleted oil reservoirs. EOR's history has made the overall challenges of the permanent sequestration of fluids well understood. The oil industry uses well-developed reservoir simulation models with computer programs that have sufficiently sophisticated computational power to routinely characterize subsurface oil reservoirs. It uses these tools extensively for oil production forecasting and to predict the state of fluids in the reservoirs, such as pressure distribution profiles and fluid flow characteristics. Oil exploration companies often conduct seismic surveys to determine the size and shape of subsurface reservoirs. They use well logging and sampling to determine the porosity, permeability, and resistivity of reservoirs and the hydrocarbons

---

[23] Thermodynamic efficiency refers to the ratio of the thermodynamic minimum energy requirement to the actual amount of energy used in the process (Zeman 2007).

[24] Conservative estimates of the potential to store CO_2 emissions geologically in North America range from 3,300 to 12,600 gigatons—that is, enough to store the CO_2 output of several coal-fired power plants for many decades.

they contain.[25] Recently published reports show that the private sector, universities, and national laboratories are developing and using computational techniques to model and simulate CO_2 injection, transport, and storage (CMI 2010; Grimstad et al. 2009; Hao et al. 2009; MacMinn and Juanes 2009; Stauffer et al. 2009).

While advances in this area are notable, further research is needed to improve the existing technologies. What is known about CO_2 injection for enhanced oil recovery could help in identifying deep underground saline formations suitable for permanent CO_2 sequestration. Carbon dioxide sequestration is being researched for its feasibility in large-scale demonstrations. Several worldwide projects are sequestering CO_2 in underground reservoirs to accelerate mainstream CO_2 mitigation.[26]

While the technology behind CO_2 injection is well developed, an integrated direct air capture and sequestration system has not been demonstrated. Furthermore, geologic sequestration of CO_2 has not been practiced on the large scale envisioned by climate engineering.

3.1.1.3 Cost factors

Cost estimates for direct air capture are based largely on theoretical calculations or assumptions, with some studies making qualitative cost comparisons (Royal Society 2009). Direct air capture's relatively high cost results from the extremely low concentration of CO_2 in the atmosphere (about 390 ppm) compared to a coal-fired stack (about 120,000 ppm).[27] Studies have reported that the steps in selective CO_2 capture and release from a solvent consume more energy—and therefore account for the majority of the costs—than transportation and underground sequestration. Besides the energy costs, other factors include transportation and logistics for sequestration of captured CO_2 and the long-term management of the sequestration site—for example, CO_2 injection, measuring, monitoring, and verification.

Cost estimates for air capture in the literature vary substantially, from a low range of $27–$135 per ton of CO_2 removed (Pielke 2009) to a higher range of $420–$630 or more per ton of CO_2 removed (Ranjan and Herzog 2010).[28] The cost estimate from Ranjan and Herzog took thermodynamics into account, concluding that direct air capture is unlikely to be a serious option in the absence of a carbon market. The literature estimates costs related to CO_2 injection and monitoring of $0.20–$30 per ton of CO_2 sequestered, reflecting a wide range of geologic parameters that could affect cost at specific

---

[25] Well logging is the process of measuring and recording the rock and fluid properties of geologic formations through drilled boreholes. It is common in the oil and gas industry for helping to find potential reservoirs, as well for gathering data to support geotechnical studies. Resistivity is a characteristic electrical property of materials defined as the electrical resistance of a conductor of unit cross-sectional area and unit length.

[26] The Department of Energy and NETL lead the federal agencies in supporting carbon capture and sequestration (CCS) research and field demonstrations. The coal-fired Mountaineer Power Plant, run by American Electric Power in West Virginia, has conducted a one-of-a-kind small-scale CCS demonstration that integrated CO_2 capture from the flue stack, injecting the CO_2 into an underground formation at the plant site. Also, the Sleipner project, run by Statoil of Norway, sequesters approximately 1 megaton of CO_2 per year in a deep saline aquifer.

[27] Direct air capture of CO_2 is expected to cost more than CO_2 capture from the flue stack of a coal-fired power plant where CO_2 concentration is substantially higher (Ranjan and Herzog 2010). Engineers from American Electric Power indicated that the present cost of capturing CO_2 from a flue stack is estimated at about $50 per ton in contrast to the likely high cost of direct air capture.

[28] These estimates apply only to the energy costs of the process. Adding capital and operations costs would increase them significantly (Ranjan 2010).

locations (Metz et al. 2005). The potentially high cost of direct air capture of CO_2 and the lack of a carbon market could impede its large-scale adoption.

3.1.1.4 Potential consequences

While direct air capture has minimally undesirable consequences (except those associated with handling process materials or chemicals), risks have been postulated for injecting large amounts of CO_2 in deep underground saline formations (Oruganti and Bryant 2009; Ehlig-Economides and Economides 2010). Experience with geologic storage is limited, and the effectiveness of risk management methods still needs to be demonstrated for use with CO_2 storage. Although CO_2 has been injected in oil reservoirs for decades, saline formations have not been proven safe or permanent. Leakage from underground sequestration sites could contaminate groundwater or cause CO_2 to escape into the atmosphere. One technical paper expressing doubt about mitigation by underground geologic CO_2 storage based its theoretical analysis on established reservoir models and assumptions of a closed form of reservoir that would render underground geologic CO_2 storage impractical and unsuitable (Ehlig-Economides and Economides 2010). These assumptions and analyses were subsequently challenged by the U.S. Department of Energy's (DOE) Pacific Northwest National Laboratory (PNNL) (Dooley and Davidson 2010).

Other studies have reported that sealing faults or fissures in an underground reservoir could cause local pressure build-up with potential rock fractures at the weakest point, in the neighborhood of a fault, and cascading problems such as well failure, CO_2 seepage, atmospheric CO_2 release, and groundwater contamination

(Oruganti and Bryant 2009).[29] Unknown or undocumented preexisting wells in the reservoir provide another way for CO_2 to escape to the atmosphere: industry experts we interviewed generally agreed that these concerns merit further analysis and a thorough characterization of geologic reservoirs.

However, studies and simulations by industry, academia, and national laboratories suggest that such risk is generally small and manageable. For example, sites are chosen for sequestration only after the thorough characterization of a reservoir and its geology. Promising sites are assessed in detail to ensure minimal or no risk. NETL's recent report advocated robust simulation to accurately model the transport and fate of CO_2 for identifying, estimating, and mitigating risks arising from CO_2 injection into the subsurface formation (Sullivan et al. 2011). Thus, CO_2 sequestration in deep underground geologic formations might be safe, provided the risks were managed adequately. Our interviews and literature review suggest that careful site characterization and appropriate monitoring and verification during injection are key to avoiding hazards, steps DOE has pursued at American Electric Power's West Virginia plant.

3.1.2 Bioenergy with CO_2 capture and sequestration

3.1.2.1 What it is

Bioenergy with CO_2 capture and sequestration (BECS) would harvest a biomass crop such as switchgrass for biofuel production and capture and sequester the CO_2 in geologic formations

---

[29] In geology, a fault is a planar fracture or discontinuity in a volume of rock, across which displacement has been significant. Large faults within Earth's crust result from the action of tectonic forces.

as it is released in the conversion of biofuel to electricity. Analogous to carbon capture and sequestration (CCS), this leverages what is known about bioenergy for fuels and CCS (Royal Society 2009).[30] As vegetation grows, photosynthesis removes large quantities of carbon from the atmosphere. A harvested crop could be used to produce biofuel or simply as a fuel to generate electricity. The CO_2 that would be released could be captured and sequestered in geologic formations. Since BECS actively absorbs CO_2 from the atmosphere over the entire life of a growing plant, this approach could, on a large scale, reduce atmospheric CO_2 (Read 2008).

3.1.2.2 Maturity and potential effectiveness

We assessed the maturity of BECS at TRL 2. Although it has been recognized that BECS can remove CO_2 from the atmosphere, it has not been applied on a scale that would affect climate change (Carbo et al. 2010). This is an emerging technology that leverages what is already known about CO_2 capture and geologic sequestration. For example, the Energy Research Center of the Netherlands has a multidisciplinary research program dedicated to BECS. BECS potentially leads to negative CO_2 emissions—that is, to CO_2 uptake from the atmosphere through natural sequestration of CO_2 in biomass (Carbo et al. 2010). Ranjan and Herzog (2010) concluded that BECS could result in negative net emissions if the biomass were harvested sustainably.

While the concept is simple, no instances of BECS are in operation. For example, BECS has not been demonstrated at any electric power generation facility. BECS is limited by the rate of growth of vegetation and conflicts with other uses of land, such as agriculture. For example, sequestering 1 gigaton of CO_2 through BECS would require more than 200,000 square miles of land for plant growth (Ranjan 2010). While BECS could benefit local environments on a small scale, the Royal Society views it as having a low to medium capacity to remove CO_2 from the atmosphere (Royal Society 2009; Royal Society 2001). According to the Royal Society, it can reduce the atmospheric CO_2 concentration by at most 50–150 ppm by the end of this century compared to a projected CO_2 concentration of 500 ppm by 2100 (Royal Society 2009).

3.1.2.3 Cost factors and potential consequences

BECS's implementation costs are variable and depend on the availability of land for harvesting biomass, unintended emissions, the targeted amount by which atmospheric CO_2 concentration would be reduced, and a carbon market, among other things (Azar et al. 2006). Other cost factors include transportation and logistics for sequestration, including the long-term management of the sequestration sites (as with direct air capture). An article by the Energy Research Center of the Netherlands concluded that incremental costs for CO_2 capture and storage are relatively low for biofuel production and are competitive with carbon capture and sequestration in fossil-fired power plants (Carbo et al. 2010). Another study reported BECS cost estimates of $150–$500 per ton of CO_2 removed and suggested that BECS looked more promising than air capture from a cost perspective, although land requirements could potentially be large

---

[30] A variant of direct air capture, CCS captures CO_2 from a fixed location such as the effluent stream of a coal-fired power plant. The large technical and scientific literature on CCS has brought it to the attention of government agencies, electric power generation corporations, and the enhanced oil recovery community (GAO 2010c; GAO 2008a). We excluded CCS from our analysis because it is not generally considered to involve deliberate modification of Earth's climate system and was therefore beyond our scope. As a forerunner of direct air capture, CCS is a key part of the bioenergy with CO_2 sequestration (BECS) method, which, at large scale, is considered to be climate engineering.

(Ranjan 2010). The literature describes BECS's technical feasibility and potential as a negative-emissions energy system that is benign and free of risks associated with some other climate engineering approaches (Read and Lermit 2005). As with direct air capture, however, the CO_2 sequestration aspects may pose risks. Furthermore, diverting resources to large-scale BECS activities could pose land-use trade-offs or affect food prices, water resources, and fertilizer use.

3.1.3 Biochar and biomass

3.1.3.1 What it is

Biochar is a carbon-rich organic material that results from heating biomass, or terrestrial vegetation, in the absence of or in a limited supply of oxygen (Whitman et al. 2010).[31] Biochar and biomass methods begin with the uptake of CO_2 in photosynthesis (Lehmann 2007). The carbon locked in plants during their growth would be converted to charcoal instead of being released to the atmosphere. Biochar differs from charcoal in that its primary use is for biosequestration rather than fuel. That is, after plants die, biochar can be buried underground or stored in soil to keep carbon from being released to the atmosphere as CO_2.

3.1.3.2 Maturity and potential effectiveness

We rated the maturity of biochar and biomass at TRL 2. Ongoing and published research is available on the sustainability of biochar to mitigate global climate change (Woolf et al. 2010). While its proof of concept has been demonstrated in published modeling and experimental results, we found uncertainties in experimental data demonstrating the efficacy of biochar as a net carbon sink. For example, how long the captured CO_2 in biochar will remain sequestered is uncertain. Similar to BECS, biochar production by pyrolysis is considered to be a carbon-negative process. Reports show its benefits to soil, but the current immaturity of biochar sequestration technology precludes it from being practiced on a scale large enough to affect the climate. Its maximum sustainable potential for reducing net CO_2, CH_4, and N_2O emissions has been estimated at 1–2 gigatons of CO_2–C equivalent per year, compared to annual anthropogenic emissions of these greenhouse gases of 15 gigatons of CO_2–C equivalent (Laird et al. 2009; Woolf et al. 2010).[32] Lehmann and colleagues (2006) quoted a higher future potential of biochar as a carbon sink of 5.5–9.5 gigatons of carbon per year by 2100. The Royal Society views biochar as low in effectiveness because its maximum anticipated reduction in atmospheric CO_2 concentration would be only 10–50 ppm by the end of this century compared to a projected atmospheric CO_2 concentration of 500 ppm in 2100 (Royal Society 2009). Therefore, biochar could be viewed as a small-scale contributor to a climate engineering approach to enhancing the global terrestrial carbon sink (Royal Society 2009).

Although producing biochar and storing it in soil have been suggested as a way to abate climate change, provide energy, and increase crop yields, scientists have expressed uncertainty about its global effect and sustainability (Woolf et al. 2010). Its emission balance is highly variable and

---

[31] Pyrolysis refers to the thermochemical decomposition of organic material at elevated temperatures in the absence of oxygen or where its supply is limited.

[32] The term CO_2–C equivalent describes the extent of global warming caused by a given type and amount of greenhouse gas, using the functionally equivalent amount or concentration of CO_2 as the reference.

largely depends on the feedstock available, the existing soil fertility, and the local energy needs (Woolf et al. 2010). While biochar and biomass sequestration methods currently represent a trivial carbon sink, experts are researching them as a means of abating climate change and improving soil fertility.

3.1.3.3 Cost factors and potential consequences

The costs of biochar and biomass are uncertain and inherently variable, depending on factors such as the type of feedstock used, the cost of pyrolysis, and carbon markets. According to one scientist, cost might depend more significantly on soil fertility outcomes. Roberts and colleagues found break-even prices of about $2–$62 per ton of CO_2 removed, depending on the pyrolysis feedstock used (Roberts et al. 2010). While the literature has reported no negative consequences of biochar or biomass in soil, their handling and application might pose safety and health risks not yet adequately managed and captured in an overall cost structure of biochar systems. Pyrolysis could also affect health and safety. Biochar's effects on emissions of N_2O, CH_4, and CO_2 from soil are poorly characterized and need to be further researched (Whitman et al. 2010). Land-use trade-offs are possible (food versus the growth of biomass for fuel), but it is unclear whether they would be a factor for biochar. For example, the sustainable potential for biochar calculated by Woolf et al (2010) assumed no land-use trade-offs.

3.1.4 Land-use management

3.1.4.1 What it is

Land-use management would enhance CO_2 uptake in trees, soils, and biomass to increase

their sequestration of carbon (DOE 2006). Although it could involve a variety of activities, we restricted our review to practices related to forestry, including reforestation, afforestation, and reductions in deforestation. Reforestation would plant trees where forests were previously cleared or burned; afforestation would plant trees where they had not historically grown. Reductions in deforestation would conserve existing forests.

3.1.4.2 Maturity and potential effectiveness

We assessed the maturity of land-use management for climate engineering at TRL 2 because of the absence of experiments demonstrating its effectiveness at the scale required to affect the climate, despite the existence of technologies and knowledge required to sequester carbon through land-use management for mitigation.[33] Bottom-up regional studies and global top-down models yield estimates of the potential for CO_2 uptake through land-use management of 1.3–13.8 gigatons of CO_2 per year in 2030 (Nabuurs et al. 2007).[34]

The effectiveness of land-use management would depend on many factors, such as the vegetation's species, location, and growth phase. For example, in the United States, afforestation could potentially sequester 2.2–9.5 metric tons of CO_2 per acre per year, reforestation 1.1–7.7 metric tons of CO_2 per acre per year, depending on the types of trees and where they were planted (Murray et al. 2005). Nabuurs and colleagues

---

[33] China has recently accomplished afforestation on a large scale for reasons unrelated to global climate change mitigation.

[34] Emissions pricing can provide financial incentives for carbon sequestration. This range of estimates of the global economic potential of land-use management assumes a price of $100 per ton of CO_2 sequestered.

reported a range for both of 0.4–14.2 tons of CO_2 per acre per year worldwide.[35] The rate of carbon accumulation also varies over a tree's life cycle, starting out slowly when a tree is first planted, then increasing. Although land-use management practices are well understood and well established, their sequestration potential could be enhanced if scientists were to improve the understanding of carbon uptake and transfer in plants and soils.

The capacity for sequestration through afforestation or reforestation also depends on the amount of land available. The estimates of sequestration potential reported by Nabuurs and colleagues suggest that the land area required to store a gigaton of CO_2 per year could range from about 100,000 to 3.9 million square miles. Other potential challenges to land-use management for climate engineering include threats to permanence, such as fire, insect outbreaks, drought, or harvesting and problems in reliably measuring, monitoring, and verifying the amount of carbon stored, although progress has been made in this area, and costs may decline further as new methods are developed (Royal Society 2009; Sohngen 2009; Canadell and Raupach 2008; Tavoni et al. 2007; Royal Society 2001).[36] Climate change itself could also affect the capacity for sequestration through land-use management, but it is unclear whether capacity would be enhanced or diminished (Nabuurs et al. 2007).

---

[35] Nabuurs and colleagues described trade-offs that could affect net sequestration from land-use management. For example, a moratorium on timber harvesting could increase the carbon sequestered in forests but could also result in the substitution of energy-intensive building materials, such as cement or concrete, for wood in the construction of buildings (Nabuurs et al. 2007).

[36] One expert noted that natural disturbances might not significantly challenge carbon sequestration through land-use management in the long term.

3.1.4.3 Cost factors and potential consequences

The costs of sequestration through land-use management would depend on a number of factors, most importantly the value of land in other uses (Sohngen 2009; Jepma 2008; Nabuurs et al. 2007; Sohngen and Sedjo 2006). Costs would also arise from implementing and managing forestry practices (such as planting seedlings or harvesting); measuring, monitoring, and verification; engaging in other transactions (for example, developing and implementing long-term sequestration contracts); and system-wide adjustments (for example, changes in the price of land) (Sohngen 2009). Although land-use management is not generally regarded as risky, some practices could affect other systems as well as climate—for example, afforestation could reduce water runoff and affect the ecology.

3.1.5 Enhanced weathering

3.1.5.1 What it is

Weathering refers to the physical or chemical breakdown of Earth's minerals in direct contact with the atmosphere. Thousands of years of the weathering of silicate rocks, for example, have removed CO_2 naturally from the atmosphere, as the CO_2 has reacted chemically with silicate rocks to form solid carbonates. The reaction can be written

$$CaSiO_3 + CO_2 \rightarrow CaCO_3 + SiO_2$$

This natural weathering of rocks could be enhanced by chemically reacting the silicate or carbonate rocks with CO_2 in the presence of sea water to produce a carbonic acid solution that could be spread in the ocean (Rau et al. 2007; Royal Society 2009).[37]

3.1.5.2 Maturity and potential effectiveness

We assessed the maturity of enhanced weathering at TRL 2. While the basic principles of enhanced weathering have been observed and a concept proposed, we did not find published experimental results describing this approach as a CO_2 reducing strategy. Neither enhanced weathering's potential nor its technological elements have been clarified. The chemical reaction that facilitates it sometimes converts silicate rocks to carbonates by reaction with CO_2. The carbonate materials resulting from enhanced weathering can be stored in the deep ocean or in soil. Similarly, the CO_2 could react with carbonate rocks in seawater for conversion and storage as bicarbonate ions in the ocean where a large pool of such ions is already present. Since Earth's silicate minerals are abundant, fixation in carbonate rocks could remove large amounts of CO_2 from the atmosphere. Scientists have made a number of proposals to hasten natural weathering.[38] For example, Rau and colleagues have reported its potential effectiveness based on models (Rau et al. 2007). While a very large potential for carbon storage in soils and oceans has been reported for this technology, its effectiveness remains uncertain. Enhanced weathering has not been practiced on a scale that would affect climate.

3.1.5.3 Cost factors and potential consequences

Enhanced weathering's costs are uncertain but are likely to be driven by mining and transportation costs (Royal Society 2009). Cost factors would include, for example, the design and implementation of a silicate-based weathering scheme and the distribution and delivery of raw materials. Rau and colleagues reported variability in cost estimates of $4–$65 per ton of CO_2 removed under various assumptions, whereas IPCC's estimate was $50–$100 per ton of CO_2 captured (Rau et al. 2007; Metz et al. 2005). Overall, this technology is expected to be relatively simple and low in cost. Enhanced weathering that entailed large-scale mining and transportation could require additional energy and water and might adversely affect air and water quality (consistent with mining activities) and aquatic life in the long term (Royal Society 2009). Viability would depend on carbon markets.

3.1.6 Ocean fertilization

3.1.6.1 What it is

Ocean fertilization releases iron to certain areas of the ocean surface to increase phytoplankton growth and promote CO_2 fixation (Buesseler et al. 2008a). Oceans act as a large sink of CO_2. Atmospheric CO_2 is exchanged at the surface and slowly transferred to deeper waters with the capacity to store about 35,000 gigatons of carbon (Royal Society 2009).[39] Phytoplankton, algae, and other microscopic plants on the ocean surface absorb CO_2 in photosynthesis and recycle it to the bottom as organic matter.

---

[37] Enhanced weathering of silicate and carbonate rocks can be represented by $CaSiO_3 + 2CO_2 + H_2O \rightarrow Ca^{2+} + 2HCO_3^- + SiO_2$ and $CaCO_3 + CO_2 + H_2O \rightarrow Ca^{2+} + 2HCO_3^-$

[38] One proposal would spread crushed olivine, a type of silicate rock, on agricultural and forested lands to sequester CO_2 and improve soil quality (Schuiling and Krijgsman 2006). Another proposal would cause the CO_2 emissions from a power plant to react with crushed limestone (mainly calcium carbonate) in the presence of seawater to spontaneously produce calcium bicarbonate ions (Rau et al. 2007).

[39] This represents a substantially large storage capacity compared to the total cumulative anthropogenic carbon additions to oceans of about 100 gigatons since preindustrial times.

As the material settles into the deep ocean bottom, the microorganisms residing there use it for food, transferring CO_2 back to the ocean as they breathe. The combined phytoplankton photosynthesis at the surface and respiration removes CO_2 at the surface and releases it at greater depths. This is called the biological pump; studies suggest manipulating this pump to expedite CO_2 sequestration.

3.1.6.2 Maturity and potential effectiveness

We assessed the maturity of ocean fertilization at TRL 2. Basic principles have been observed and reported, and the concept has been formulated, with multiple studies proposing iron fertilization as an option for reducing CO_2 in the atmosphere. Oceans are the largest natural absorbers of CO_2 on the planet (at about 337 gigatons of CO_2 per year) and the largest natural reservoir of excess carbon (Rau 2009). However, most of the CO_2 the oceans absorb is released back to the atmosphere in a continuous exchange while only a small portion of it is transferred to and sequestered in the deep ocean.

The large number of theoretical studies attempting to understand fertilization's complexities with sophisticated ocean models—as many as 12 between 1993 and 2008—have been complemented with only a few small-scale field experiments, whose results were uncertain and not well documented. Ocean fertilization studies suggest that 30,000–110,000 tons of carbon could be sequestered from air by adding 1 ton of iron to certain parts of the ocean, but verifying this technology's effectiveness is difficult and uncertain (Buesseler et al. 2008b).[40] For example, modeling simulations suggest a cumulative

storage potential of 26–70 gigatons of carbon (equivalent to 95–255 gigatons of CO_2) for large-scale ocean fertilization—relatively low compared to terrestrial sequestration potential in vegetation (200 gigatons of carbon) or in deep geological formation (several hundred gigatons of carbon) (Bertram 2009).

Another study based on models reported that large-scale sustained iron fertilization (30 percent of the global ocean area) could store at most 0.5 gigatons of carbon (equivalent to about 2 gigatons of CO_2) per year. This amount is small compared to anthropogenic emissions of approximately 8–9 gigatons of carbon (equivalent to about 30–33 gigatons of CO_2) per year. According to the Royal Society, ocean fertilization could reduce the atmospheric CO_2 concentration by a maximum of 10–30 ppm by the end of this century, which would be considered to be low in effectiveness. While these estimates have not been substantiated experimentally, these studies show that even sustained fertilization of oceans would have only a minor effect on the increasing atmospheric CO_2 concentration (Secretariat of the Convention on Biological Diversity 2009).

Ocean fertilization as a long-term carbon storage strategy has not been demonstrated (Buesseler et al. 2008b). The literature characterizes its effectiveness as highly uncertain, the models governing biochemical cycling of nutrients and the circulation of ocean currents as poorly understood or uncertain, and the strategy for mitigating CO_2 as risky. For example, the science is unclear regarding ecological consequences, the duration of carbon sequestered in the oceans, and the frequency with which iron should be added (Buesseler et al. 2008b). Scientists are researching the ocean's biochemical processes and the effects and efficacy of iron fertilization to better understand them.

---

[40] One ton of carbon corresponds to 3.67 tons of CO_2.

3.1.6.3 Cost factors and potential consequences

Ocean fertilization could be cost-effective at capturing and sequestering atmospheric CO_2 in the deep ocean, but relatively little is known about its efficacy.[41] The design and implementation of any ocean fertilization scheme, including mining, distribution, and delivery of materials, would affect its success. The literature has reported significant uncertainty with respect to cost. Some ocean fertilization modeling has helped determine its efficiency at removing carbon from the atmosphere but estimating a cost range is difficult. One estimate put the minimum cost at approximately $8 per ton of CO_2 removed (Buesseler et al. 2008b). An evaluation by Boyd characterized ocean fertilization as a medium-risk strategy with costs of $8–$80 per ton of CO_2 removed (Boyd 2008).

Because ocean fertilization is not well understood and is largely theoretical, it could pose ecological risks (Royal Society 2009). A report from the Woods Hole Oceanographic Institution indicated that iron-fertilized phytoplankton blooms could eventually prevent oceans from sustaining life. An image in that report showed bloom and anoxic (or dead) zones stretching for hundreds of kilometers (Buesseler et al. 2008b).[42] The Royal Society and the U.K. House of Commons Science and Technology Committee reported that ecosystem-based methods—whether fertilizing the ocean or blocking sunlight—would be subject to unknown risks if implemented on a large scale. Other studies have also presented images of the unintended consequences of

manipulating ecosystems—dead zones in the sea resulting from phytoplankton boom are an example (Buesseler et al. 2008b). Other potential risks of ocean fertilization are greater ocean acidification, additional emissions of greenhouse gases, and the reduction of oxygen in the ocean to levels some species cannot tolerate (Buesseler et al. 2008b).

3.2 Selected SRM Technologies

In this section, we summarize our assessment of the maturity of selected SRM technologies and present information from peer-reviewed literature on their potential effectiveness, cost factors, and potential consequences. TRL ratings indicate the maturity of each technology. Potential effectiveness is described in terms of the anticipated ability to counteract warming caused by doubling the preindustrial atmospheric concentration of CO_2. In calculating our ratings, we relied on reported results from

- climate engineering modeling studies using general circulation models (GCM) and

- energy balance studies of the effects of increasing reflectivities.

Cost factors represent resources required to counteract global warming from doubling the preindustrial atmospheric concentration of CO_2 or, for technologies that are not anticipated to be fully effective, the resources required to counteract warming to the maximum extent possible. Potential consequences associated with each technology include reported negative consequences and cobenefits. (See table 3.2.)

---

[41] Despite the fact that oceans exchange large quantities of CO_2 with the atmosphere in a natural process, comparatively little is known about sequestering CO_2 by ocean fertilization.

[42] Anoxia means the absence of oxygen. Algal blooms in the ocean can deplete available oxygen in the water, leading to dead or anoxic zones.

Table 3.2 (continued)

Technology	Maturity[a]	Potential effectiveness[b]	Cost factors[c]	Potential consequences[d]
Stratospheric aerosols	**Low (TRL 1):** • Basic principles understood and reported • No system concept proposed	**Potentially fully effective:** • Aerosols must be continuously replaced	• Design, fabrication, testing, acquisition, and deployment of aerosol delivery scheme, including distribution and delivery mechanisms, fabrication of aerosol dispersal equipment, and all associated infrastructure • Literature-based estimates vary significantly: $35 billion to $65 billion in the first year; $13 billion to $25 billion in operating cost each year thereafter	• Little change in global average annual precipitation • Disruption of Asian and African summer monsoons with accompanying reduction in precipitation • Delayed ozone layer recovery in southern hemisphere and about a 30-year delay in recovery of Antarctic ozone hole • Scattering interference with terrestrial astronomy • Efficiency of solar-collector power plants reduced by increased diffuse radiation
Marine cloud brightening	**Low (TRL 2):** • Basic principles understood and reported • System concept proposed • Proof of concept not demonstrated	**Potentially fully effective:** • Model-dependent estimates of effectiveness vary • Clouds must be continuously brightened	• Design, fabrication, testing, acquisition, and deployment of a fleet of 1,500 wind-driven spray vessels • Fleet infrastructure and operation • Estimates in the scientific literature vary significantly at $42 million for development, $47 million for production tooling, $2.3 billion to $4.7 billion for 1,500-vessel fleet acquisition	• Small changes in global average temperature, regional temperatures, and global precipitation • Large regional changes in precipitation, evaporation, and runoff; both precipitation and runoff increase, and the net result might not "dry out" the continents

Table 3.2 Selected SRM technologies, continues on next page

TECHNOLOGY ASSESSMENT GAO-11-71 31

Technology	Maturity[a]	Potential effectiveness[b]	Cost factors[c]	Potential consequences[d]
Scatterers or reflectors in space • Earth orbit • Deep space	**Low (TRL 2):** • Basic principles understood and reported • System concepts proposed, but proof of concept not demonstrated	**Potentially fully effective:** • Spacecraft's limited lifetime	• Design, fabrication, testing, acquisition, and deployment of a fleet of millions to trillions of reflecting or scattering spacecraft • Launch vehicle • Infrastructure and operation • Estimates in the scientific literature vary significantly: an estimate of $1.3 trillion and an estimate of less than $5 trillion	**Earth-orbit technologies:** • A cool band in the tropics with unknown effects on ocean currents, temperature, precipitation, and wind • A multitude of bright "stars" in the morning and evening that would interfere with terrestrial astronomy **Deep-space technologies:** • Annual average tropical temperatures a little cooler • Annual average higher latitude temperatures a little warmer • Small reduction of annual global precipitation
Terrestrial reflectivity • Deserts • Flora • Urban or settled areas	**Low (Up to TRL 2):** • Basic principles understood and reported • One technology proposed a system concept but without demonstrated proof of concept	**Potential effectiveness of 0.21 (urban areas) to more than 57 percent (deserts)** • Sustainability issues: maintaining reflectivity and missing information on reflective flora	• Design, fabrication, testing, acquisition, and deployment of reflective material or flora • Infrastructure and maintenance • Estimates in the scientific literature to maintain reflectivity vary greatly from $78 billion (urban areas) to $3 trillion per year (deserts)	• Cool deserts might change large-scale patterns of atmospheric circulation • Reflective crops would probably not significantly affect global average temperature but might reduce regional summer temperatures • Reflective urban areas would probably not affect global average temperature but might reduce air-conditioning costs

Table 3.2 Selected SRM technologies: Their maturity and a summary of available information. Source: GAO.

[a] In this report, we considered each technology's maturity in terms of its readiness for application in a system designed to address global climate change. To do this, we used technology readiness levels (TRL), a standard tool that some federal agencies use to assess the maturity of emerging technologies. We characterized technologies with TRL scores lower than 6 as "immature" (see section 8.1). The TRL rating methodology considers the maturity level of the whole integrated system rather than individual components of a particular technology.

[b] We assessed potential effectiveness in terms of a technology's potential ability to counteract global warming caused by doubling the preindustrial CO_2 concentration.

[c] Cost factors are resources a system uses to counteract global warming caused by doubled preindustrial atmospheric CO_2 concentration, or for technologies that are potentially not fully effective, resources required to counteract global warming to the maximum extent possible. Some of the studies we reviewed indicate possible cost levels, which we provide here for illustration. We did not evaluate this information independently.

[d] Includes potential consequences, risks, and cobenefits.

3.2.1 Stratospheric aerosols

3.2.1.1 What it is

Deploying aerosols would use knowledge gained from volcanic eruptions that inject aerosols into the stratosphere, cooling Earth for short periods. Aerosols smaller than 1 micrometer in diameter (1 millionth of a meter) would cool Earth primarily by scattering a fraction of the solar radiation. While enough solar radiation would be scattered back into space to cool Earth, a larger fraction would be scattered toward Earth, increasing diffuse radiation (Robock 2000). Larger aerosols would scatter solar radiation less efficiently and absorb both solar and thermal radiation, acting somewhat like a greenhouse gas (Rasch, Crutzen, and Coleman 2008; Rasch, Tilmes et al. 2008). If the volcanic sulfate aerosols were sufficient to cool Earth, the sulfates would accumulate in size and remain in the stratosphere for about 1 year.[43]

3.2.1.2 Maturity and potential effectiveness

We assessed stratospheric aerosol technology at TRL 1 because only basic principles have been reported. We could not rate this technology at TRL 2 because we found no system concepts reported in the literature. Recent estimates using complex coupled atmosphere-ocean general circulation models indicated that about 3 million tons of sulfur injected per year into the stratosphere and forming volcanic-sized sulfate aerosols would compensate for the doubled CO_2 concentration (Rasch, Crutzen, and Coleman 2008). In a recent investigation using a chemistry

climate model, Heckendorn and colleagues found that sulfates from continuous injection of sulfur gas formed larger aerosols that would be less effective than volcanic sized aerosols (Heckendorn et al. 2009). Because sulfate aerosols have a lifetime of about a year in the stratosphere, they must be replenished to sustain their cooling effect (Rasch, Crutzen, and Coleman 2008).

3.2.1.3 Cost factors and potential consequences

It could cost $35 billion to $65 billion in the first year and $13 billion to $25 billion in each subsequent year to inject sufficient sulfate aerosols into the stratosphere to counteract global warming caused by doubling preindustrial CO_2 concentration. Robock and colleagues estimated the cost of injecting 1 million tons of a sulfur gas (that will become sulfate aerosols) per year into the stratosphere (Robock et al. 2009). Since about 3 million tons of sulfur might be required to counteract global warming caused by doubling preindustrial CO_2 concentration, we scaled Robock and colleagues' cost estimate, assuming no economy of scale, to 3.2 million and 6 million tons per year of hydrogen sulfide and sulfur dioxide, respectively (gases containing 3 million tons of sulfur). The scaled cost estimate is $35 billion to $65 billion in the first year (the cost of the airplanes used to inject the aerosols plus 1 year of operations) and $13 billion to $25 billion in operating costs in each subsequent year to sustain the effort. Robock and colleagues considered several potential aerosol injection systems, including KC-135 aircraft-refueling tankers and F-15 aircraft. They found that the total cost of using the aircraft-refueling tankers would be lower than the total cost of the alternatives, but the tankers do not fly high enough (Robock et al. 2009).

---

[43] While the published research has focused on sulfate aerosols (Royal Society 2009), other aerosols such as alumina (Teller et al. 1997) and self-levitated nanoparticles (Keith 2010) have also been considered.

Using the F-15s was the least expensive among the remaining alternatives. Robock and colleagues' estimated operating cost for the F-15s was an upper bound based on the hourly cost of the tankers; the authors expected that the hourly cost of operating F-15s would be lower because they use less fuel and fewer pilots than the tankers. However, because the F-15s are smaller than the tankers, they would require more than ten times the number of trips that the tankers would require to inject the same quantity of aerosols. Other alternatives considered by the authors, including injection systems based on artillery or balloons, would be significantly more expensive than the fighter aircraft. The scaled estimates do not include system design, fabricating aerosol dispersal equipment, or infrastructure.

Volcanic stratospheric sulfate aerosols increase diffuse solar radiation, which can increase the growth of terrestrial vegetation (Robock et al. 2009). Cooling by these aerosols can interfere with the hydrological cycle (Trenberth and Dai 2007). The surface area of these aerosols can lead to reactions that deplete stratospheric ozone (Tilmes et al. 2008; Solomon 1999). Robock and colleagues reported performing a modeling study using an IPCC "business-as-usual" scenario with an increase in greenhouse gases and sufficient stratospheric sulfate aerosols to significantly cool Earth. They found little annual average change in global precipitation but significantly reduced precipitation in India, with large reductions in summer monsoon precipitation in India and northern China that could threaten food and water supplies. They found a similar reduction in the Sahel in Africa. They also found that abruptly stopping the injection of aerosols would raise temperature rapidly and be difficult to adapt to.

In another modeling study using the same greenhouse gas scenario, Tilmes and colleagues found that changes in stratospheric dynamics and chemistry delayed the recovery of the ozone layer in middle and high latitudes in the southern hemisphere and reduced the ozone layer in high latitudes in the northern hemisphere (Tilmes et al. 2009). The recovery of the Antarctic ozone hole would be delayed by about 30 years. They stated that the increase in ultraviolet radiation of up to 10 percent observed in the middle and high latitudes in the 1980s and 1990s would probably worsen.

Using an aerosol-chemistry climate model, Heckendorn and colleagues found larger sulfate aerosols, which increased stratospheric water vapor and reduced stratospheric ozone (Heckendorn et al. 2009). Additional water vapor (a greenhouse gas) would reduce effectiveness but reduced ozone (another greenhouse gas) would increase effectiveness. The net effect is not known because detailed radiation forcing calculations were beyond the scope of the 2009 study.

Other collateral consequences of stratospheric aerosols would include negative effects on astronomy and on solar energy power plants. Suspended above all terrestrial telescopes, stratospheric aerosols would interfere with terrestrial optical astronomy. Scattering from stratospheric aerosols would also reduce the efficiency of power plants that concentrate solar radiation to generate electricity. Although solar radiation scattered from aerosols would result in significant diffuse radiation, the concentrators in these power plants cannot use it. For example, the peak power output of Solar Electric Generating Stations in California fell up to 20 percent after Mount Pinatubo erupted, even though total solar radiation was reduced by less than 3 percent (Murphy 2009). Aerosol effects in the stratosphere could be reversed by stopping their injection because sulfate aerosols remain in the stratosphere for approximately 1 year.

3.2.2 Cloud brightening

3.2.2.1 What it is

Reflectivity in clouds generally increases as the number of water droplets in them increases (Twomey 1977). Latham and colleagues proposed to increase the reflectivity of marine clouds by increasing the number of water droplets (Latham et al. 2008). They proposed to loft droplets of sea water micrometers in diameter that would shrink by evaporation as they rose into the base of the clouds, where moisture would condense, and increase their number (Latham et al. 2008). In designing wind-driven spray vessel-based cloud brightening equipment, Salter, Sortino, and Latham (2008) proposed to avoid the problems of remotely operating and maintaining sails, ropes, and reefing gear by using Flettner rotors—vertical spinning cylinders that produce forces perpendicular to the wind direction—instead of sails.

3.2.2.2 Maturity and potential effectiveness

We assessed cloud brightening technology at TRL 2. Basic principles have been reported, allowing at least TRL 1. Demonstration of proof of concept has not been reported (Salter, Sortino, and Latham 2008), ruling out TRL 3. A system concept has been proposed, and there is encouraging evidence that this technology might work: Ship tracks (which are white streaks observed in satellite images of the oceans that are attributed to sulfate aerosols in the exhaust trails from ships) indicate that adding aerosols to the marine environment can make clouds, but they fall short of proof of concept that lofting droplets of sea water into marine clouds will brighten them as assumed in the analyses discussed below. Having a system concept does not automatically qualify this technology for TRL 2 but it cannot

be ruled out, given the information available in Salter, Sortino, and Latham (2008).

Four recent investigations of cloud brightening reported effectiveness ranging from fully effective to fully effective with a significant margin. Latham and colleagues used two different atmosphere-only general circulation models and calculated the increased reflectivity of brightened clouds. They found full effectiveness with significant margin for one when they brightened all marine clouds and full effectiveness for the other when they brightened clouds over 35 to 45 percent of the ocean area (Latham et al. 2008). Using analytical methods, Lenton and Vaughan found full effectiveness but warned that conversion of droplets reaching the base of the clouds into droplets in the clouds is not well understood and, if the conversion is insufficient, this technology would not be effective (Lenton and Vaughan 2009). Rasch and colleagues used a fully coupled atmosphere-ocean general circulation model and found full effectiveness if clouds were brightened over between 40 percent and 70 percent of the oceans (Rasch et al. 2009). Bala and colleagues used a similar atmosphere model coupled to a simple slab-ocean/sea-ice general circulation model and found full effectiveness when they reduced water droplet size in all marine clouds (Bala et al. 2010). Brightened clouds have a lifetime of a few days and must be continuously brightened to sustain cooling (Latham et al. 2008).

3.2.2.3 Cost factors and potential consequences

It could cost $2.4 billion to $4.8 billion to brighten enough marine clouds to compensate for a doubling of the concentration of CO_2 in the atmosphere. Salter, Sortino, and Latham (2008) estimated that full effectiveness would require a fleet of 1,500 of their wind-driven spray vessels.[44] Their cost estimate did not include system testing, acquisition, deployment, infrastructure, and operation.

The investigations using coupled atmosphere-ocean general circulation models predicted climate changes. Rasch, Latham, and Chen (2009), using a fully coupled ocean-atmosphere model, found that as they brightened increasing fractions of clouds, they not only could counteract global warming caused by doubling atmospheric CO_2 concentrations but could also counteract the effects of this warming on sea ice and precipitation—but not all at the same time. For example, when they counteracted global warming, they overcompensated for the loss of south polar sea ice and the change in global precipitation and undercompensated for the loss of north polar sea ice (Rasch, Latham, and Chen 2009). Bala and colleagues used an atmosphere coupled to a simple slab-ocean/sea-ice model and found that

- changes in global and regional annual average temperatures were small,

- changes in global annual precipitation were small, and

- regional changes in precipitation, evaporation, and runoff were large. Precipitation and runoff increased over land, particularly over Central America, the Amazon, India, and the Sahel, suggesting that this technology might not dry the continents (Bala et al. 2010).

---

[44] The total rough cost estimate for the cloud brightening system would be $2.4 billion to $4.8 billion. This cost estimate is made up of Salter, Sortino, and Latham's estimates of $3.1 million for the first 2 years of engineering; $39 million for the next 3 years for final design, including construction of a prototype; $47 million for production tooling; and production costs of $2.3 billion to $4.7 billion ($1.56 million to $3.13 million each) for 1,500 45-meter, 300-ton wind-driven spray vessels (Salter et al. 2008).

The brightness of clouds could be returned to normal within a few days of ceasing to deploy the cloud brightening technology (Latham et al. 2008).

3.2.3 Scatterers or reflectors in space

3.2.3.1 What it is

Proposals have been made to reduce the solar radiation that reaches Earth by placing scatterers or reflectors in Earth orbit or in deeper space at a stable position between Earth and the Sun called the inner Lagrange point (or L1)—approximately 1 percent of the distance from Earth toward the Sun—where gravitational and orbital forces are balanced.

Proposed technologies include scatterers or reflectors in Earth orbit. NAS dismissed the use of 55,000 110-ton 100-square kilometer reflective solar "sails" in orbit that would reflect 1 percent of solar radiation as "a very difficult if not unmanageable control problem" (NAS 1992).[45]

Pearson, Oldson, and Levin (2006) proposed Saturn-like rings of space dust or parasol spacecraft. To be practical, the space dust option would require the ability to fabricate in space. The ring of spacecraft would consist of 5 million parasol spacecraft, each measuring 5 km long by 200 m wide (1 square km) and having mass of 1,000 kg. They would be electromagnetically tethered in Earth's equatorial plane at altitudes between 1,300 km and 3,200 km. The spacecraft's parasols would point at the Sun and shade the tropics of the winter hemisphere.

---

[45] In 1992, reflecting 1 percent of solar radiation was thought to counteract the global warming from doubling the concentration of CO_2 in the atmosphere (NAS 1992). Up-to-date modeling studies indicate that reflection of about 1.8 percent is required (Govindasamy and Caldeira 2000; Govindasamy et al. 2002; Caldeira and Wood 2008).

The proposed options also included scatterers or reflectors at L1:

- a 3,400-ton, 1,800-km diameter diaphanous scattering screen fabricated in low Earth orbit (Teller et al. 1997);

- a 100-million ton, 2,000-km diameter, 10 micrometer thick opaque disc or transparent prism made from moon glass (Early 1989);

- a 420-million ton, 3,600-km diameter, 5.1 micrometer thick iron mirror made from asteroids (McInnes 2002);

- 16 trillion spacecraft (a total of 19 million tons), each 0.6 meters in diameter and 5 micrometers thick, covering an ellipse 6,200 km by 7,200 km (Angel 2006).

The first three technologies are impractical at this time because they require manufacturing capabilities in space. The fourth technology would consist of 16 trillion autonomous fliers, manufactured on Earth, launched electromagnetically into orbit, and moved into position with ion propulsion. Once in position, they would use a system analogous to the global positioning system and radiation pressure motive power with tilting mirrors for station-keeping.

3.2.3.2 Maturity and potential effectiveness

We assessed scattering or reflecting technologies in space at TRL 2. Basic principles have been reported and system concepts have been proposed, allowing at least TRL 1, but demonstration of proofs of concept have not been reported (Angel 2006; Pearson et al. 2006), ruling out TRL 3. Having system concepts does not automatically qualify these technologies for TRL 2 but it cannot be ruled out, given the

information available in Angel (2006) and in Pearson, Oldson, and Levin (2006).

Pearson, Oldson, and Levin (2006) used a simplified one-dimensional energy balance model to design a system of parasol spacecraft to reduce solar radiation to compensate for doubled preindustrial CO_2 concentrations in the atmosphere. Their design study indicated that this could be accomplished by shading about 36 percent of a Saturn-like equatorial ring with their parasol spacecraft.

Angel's autonomous spacecraft fliers were designed to reduce solar radiation by the 1.8 percent required by general circulation models (in this case, an atmospheric general circulation model coupled to slab ocean and sea-ice models) (Govindasamy and Caldeira 2000) to compensate for global warming caused by doubling the preindustrial CO_2 concentration.

None of these space-based SRM technologies would be a realistic contributor in the short term. They should not be dismissed from future consideration, particularly if climate engineering were to be employed for as long as a century (Royal Society 2009). However, the spacecraft would have to be replaced when they reached the end of their service life to sustain cooling.

3.2.3.3 Cost factors and potential consequences

Following NAS's assertion that the cost of establishing space-based climate engineering projects would be dominated by launch costs (NAS 1992), Pearson, Oldson, and Levin (2006) estimated a cost of $1.3 trillion for their equatorial Saturn-ring-like collection of reflectors. Their launch cost was based on a proposed ram accelerator and an orbiting tether, achieving a low Earth orbit launch cost of $250 per kg. This cost estimate did not include design, fabrication, testing, acquisition, deployment, infrastructure, or operation. They did not provide an explicit projected lifetime for their spacecraft. However, they did explore the consequences of a 100-year lifetime. Following on their launch costs as discussed, the replacement cost estimate would be $13 billion per year.

It could cost less than $5 trillion for Angel's fliers at L1. Fabrication costs were estimated at $50 per kg, which Angel rounded up to $1 trillion. Estimates of launch costs were based on 20 electromagnetic launchers each launching 800,000 fliers into orbit every 5 minutes for 10 years. The electromagnetic launchers would put the fliers into orbit, and ion propulsion would move them to L1, where the fliers would use mirrors to adjust radiation pressure from solar radiation to maintain position. The cost estimate for the launchers was $600 billion and the estimated cost of electrical energy for the launchers was $150 billion; Angel rounded the sum to $1 trillion, corresponding to a launch cost of $50 per kg. Angel stated that a total project cost, including development and operations, of less than $5 trillion seemed possible but gave insufficient detail to evaluate development and operation costs. Also, he did not explicitly mention testing, acquisition, deployment, and infrastructure. The projected lifetime for the fliers is 50 years, which means that 320 billion fliers would have to be replaced every year, but Angel did not provide an estimated cost for replacement.

Orbital equatorial Saturn-ring-like disposition of reflectors is a regional technology that would shade and cool the winter portion of the tropics. The design study used a simplified energy balance model of Earth's climate system, not a general circulation model (GCM). Therefore, climate responses other than a set of average temperatures

for bands of latitudes are not available. The effects on the ocean currents, ocean temperature, precipitation, and wind are unknown. However, a multitude of bright "stars" at morning and evening would interfere with terrestrial astronomy.

Uniformly reducing solar radiation with reflectors or scatterers at L1 enough to counteract the warming effect of doubling the concentration of CO_2 might not significantly reduce CO_2 fertilization from doubling CO_2. Govindasamy and colleagues modeled this effect with normal and uniformly reduced solar radiation at both the concentration of CO_2 in 1991 and double the concentration of CO_2 in 1991 (Govindasamy et al. 2002). In their modeling study, they chose a reduction in solar radiation that could nearly counteract the warming effect of doubling the concentration of CO_2. They found that doubling CO_2 resulted in CO_2 fertilization—that is, plant productivity increased by 76–77 percent and biomass increased by 87–92 percent.[46] When they uniformly reduced solar radiation to counteract the warming effect of this doubling of the concentration of CO_2, they found that plant productivity fell by 2.3–3 percent and biomass fell by 1.9–4.7 percent. Govindasamy and colleagues indicated that in reality, CO_2-fertilized ecosystems might encounter nutrient limitations, diminishing the magnitude but not changing the direction of the CO_2 fertilization. Furthermore, they indicated that CO_2 fertilization might affect ecosystems in ways not represented in the model through species abundance and competition, habitat loss, biodiversity, and other disturbances. This investigation applies directly to reflectors or scatterers at L1 that uniformly reduce solar radiation without otherwise affecting the Earth system. Therefore, this modeling study indicated

[46] In this context, plant productivity is net primary productivity, which is net carbon uptake by vegetation.

that CO_2 fertilization would outweigh reduction in plant productivity because of uniformly reduced solar radiation from reflectors or scatterers at L1.

Modeling studies indicate that SRM technologies that counteract the greenhouse effect of a doubled preindustrial concentration of CO_2 by uniformly reducing solar radiation also indicate that the globally averaged engineered climate is very similar to the globally averaged preindustrial climate (Caldeira and Wood 2008; Govindasamy et al. 2002; Govindasamy and Caldeira 2000). These studies indicated that annual average tropical temperatures would be a little cooler, the higher latitudes might be a little warmer, and the reduction of annual global precipitation would be small.

Since the spacecraft in Earth orbit and at L1 would be controlled, it should be possible to reverse these technologies. It is assumed that parasol spacecraft in Earth orbit, which are controlled to maximize shading, could be reversed by commanding the parasols to minimize shading (Pearson et al. 2006). Fliers at L1 could be reversed by commanding the fliers to go into halo orbits (Angel 2006).

3.2.4 Reflective deserts, flora, and habitats

3.2.4.1 What it is

Increasing Earth's surface reflectivity in deserts, flora, and settled areas has been proposed. Gaskill would double the reflectivity of deserts by covering them with white polyethylene, estimating that up to 12 trillion square meters of Earth's deserts (about 2 percent of Earth's surface) would be suitable for reflectivity enhancement (Gaskill 2004; Gaskill n.d.).

Similarly, Ridgwell and colleagues proposed increasing the reflectivity of crops by selecting varieties that are glossy or have reflective shapes and structure (Ridgwell et al. 2009). Hamwey proposed to increase the reflectivity of open shrubland, grasslands, and savannah and to double the reflectivity of all human settlements, excluding agricultural land (Hamwey 2007). Akbari, Menon, and Rosenfeld (2009) proposed to increase the reflectivity of urban roofs and pavement.

3.2.4.2 Maturity and potential effectiveness

We assessed increased reflectivity of desert technology at TRL 2. Basic principles have been reported and a system concept has been proposed, allowing at least TRL 1, but demonstration of proof of concept has not been reported (Gaskill 2004; Gaskill n.d.), ruling out TRL 3. Having a system concept does not automatically qualify this technology for TRL 2 but it cannot be ruled out given the information available in Gaskill (2004) and Gaskill (n.d.). We assessed technologies for increasing the reflectivity of flora and settled areas at TRL 1 because only basic principles have been reported; the absence of system concepts precluded a rating of TRL 2 (Ridgwell et al. 2009; Hamwey 2007).

Technologies for increasing the reflectivity of deserts could potentially be more than 57 percent effective in compensating for global warming from doubled preindustrial CO_2. Gaskill proposed to increase reflectivity from 36 to 80 percent over 10 trillion square meters of the 12 trillion square meters of desert areas that he deemed suitable (Gaskill 2004; Gaskill n.d.). The Royal Society's (2009) and Lenton and Vaughan's (2009) interpretation of Gaskill corresponded to an effectiveness of 74 percent. Lenton and Vaughan's refinement of Gaskill's proposal corresponded to 57 percent effectiveness, accounting for lower average intensity of solar radiation over land and absorption in the atmosphere. However, they also stated that deserts have higher-than-average solar radiation because they are generally in the lower latitudes, so that increased reflectivity would be somewhat more effective (Lenton and Vaughan 2009). Sustaining reflective deserts would require maintenance.

Increasing the reflectivity of flora could be up to about 25 percent effective. Ridgwell and colleagues investigated the effect of increasing the reflectivity of crops with a fully coupled climate model (Ridgwell et al. 2009). They focused on an increase of 20 percent, asserting that an increase of 35 percent observed after coating plants with a white chalky suspension provided a first-order guide as to the possible upper limit of reflectivity increase. They found a global average cooling of only 0.11 degrees Celsius. Hamwey investigated increasing the reflectivity of open shrubland, grasslands, and savannah with a static two-dimensional radiative transfer model (Hamwey 2007). His preliminary estimate was that an increase in reflectance of 25 percent corresponded to about 16 percent effectiveness. Lenton and Vaughan interpreted these results with energy balance analyses (Lenton and Vaughan 2009). Following Ridgwell and colleagues, their estimate—using a larger area estimate and a 40 percent increase in reflectance—corresponded to an upper limit of about 9 percent effectiveness. Their interpretation of Hamwey's data corresponded to essentially the same effectiveness as Hamwey's—about 16 percent. Thus the total effectiveness of reflective flora—cropland, open shrubland, grasslands, and savannah combined, using Lenton and Vaughan's reinterpretations based on energy balance—would be up to about 25 percent.

Because crops are customarily replanted annually, no additional effort should be required to maintain their reflectivity (Ridgwell et al. 2009). Hamwey provided no information on the effort required to maintain the reflectivity of open shrubland, grasslands, and savannah (Hamwey 2007).

Increasing the reflectivity of settled areas could be about 4.3 percent effective. Akbari, Menon, and Rosenfeld's (2009) estimate for urban area equal to 1 percent of Earth's land surface and a net increase for urban reflectivity by 10 percent corresponded to an effectiveness of only about 1.2 percent. However, Lenton and Vaughan (2009) suggested that the urban area Akbari, Menon, and Rosenfeld (2009) used, could have been 5.6 times overestimated, in which case increasing the reflectivity of urban areas would be only about 0.21 percent effective. Hamwey's (2007) estimate for doubling reflectivity for areas of human settlement (not including agricultural land) corresponded to an estimated overall effectiveness of about 4.6 percent. Lenton and Vaughan's correction to Hamwey's estimate accounting for absorption in the atmosphere and an underestimate in solar radiation corresponded to an effectiveness of about 4.3 percent. Maintaining high reflectivity would be the sustainability issue for these technologies.

3.2.4.3 Cost factors and potential consequences

The maintenance cost for reflective deserts that could potentially compensate for more than 57 percent of the doubling of the concentration of CO_2 in the atmosphere could be about $3 trillion per year. Gaskill proposed to increase reflectivity of 10 trillion square meters of the deserts (Lenton and Vaughan 2009; Royal Society 2009). The Royal Society provided the following cost estimate for reflective deserts

(Royal Society 2009): if the cost of reflective sheeting, with an allowance for routine replacement from damage, were somewhat similar to that of painting, it would be several trillion dollars per year. The Royal Society's method would yield an annual maintenance cost for reflective deserts of about $3 trillion (Royal Society 2009). The estimates did not include design, fabrication, testing, acquisition, installation, or infrastructure costs.

We found no cost estimates for increasing the reflectivity of flora in the peer-reviewed literature (Royal Society 2009). We found no cost estimates for increasing the reflectivity of areas of human settlement in the peer-reviewed literature. However, the estimated maintenance cost for urban areas that would compensate for 0.21 to 1.2 percent of the doubled concentration of CO_2 in the atmosphere was from about $78 billion to about $440 billion per year. The Royal Society (2009) made a rough estimate of the costs of painting urban surfaces and structures white using standard costs for domestic and industrial painting. Assuming repainting once every 10 years, it estimated combined paint and manpower costs on the order of $0.30 per square meter per year. The urban area Akbari, Menon, and Rosenfeld (2009) studied was 1 percent of Earth's land area—that is, about 1.47 trillion square meters. Using the Royal Society's (2009) cost estimation method, maintenance would cost about $440 billion per year. Lenton and Vaughan (2009) suggested that the global urban area might be only about 260 billion square meters, in which case maintenance would cost about $78 billion per year. These estimates did not include design, fabrication, testing, acquisition, installation, or infrastructure.

Desert reflectivity is regional. The Royal Society (2009) stated that as with other very localized SRM technologies, this approach could change large-scale patterns of atmospheric circulation, like the East African monsoon that brings rain to sub-Saharan Africa. The technology could be reversed by removing the reflective material.

A 2009 modeling study by Ridgwell and colleagues indicated that increasing the reflectivity of crops by 20 percent would not create a significant effect on global average temperature but that reflective crops could have an appreciable cooling effect regionally. This study indicated that reflective crops could depress temperatures by more than 1 degree Celsius during summer months in a pattern broadly corresponding to the densest cropland coverage in the model.

Hamwey's 2007 investigation of increasing the reflectivity of open shrubland, grassland, and savannah used a radiative transfer model, and Lenton and Vaughn's 2009 investigation of reflective crops and open shrubland, grassland, and savannah used an analytical approach based on energy balance considerations, so neither investigation can be used to evaluate climate consequences other than global average temperature. Hamwey did not discuss ecological issues associated with such a massive change to natural flora. Increasing the reflectivity of flora could reduce overall photosynthesis, which could reduce net carbon uptake by vegetation and crop yields. However, this is judged to be of relatively low risk, since photosynthesis tends to be light-saturated during most of the growing season (Royal Society 2009). This technology could be reversed by replanting original flora.

Since the analyses of reflective urban areas (Akbari et al. 2009; Lenton and Vaughan 2009) and human habitats (Lenton and Vaughan 2009; Hamwey 2007) were based on analytic estimates of radiative forcing, radiative transfer, or energy balance, their results cannot be used to evaluate climate consequences other than global average temperatures. However, reflective surfaces could reduce air-conditioning costs (Levinson and Akbari 2010). Effects would be reversible by returning reflective surfaces to their original condition.

3.3 Status of knowledge and tools for understanding climate engineering

Gordon (2010, 7–8) identified 26 examples of areas of climate research that are important to understanding climate engineering and 8 examples of climate engineering research tools. The report described resources at several federal agencies that could help advance climate engineering research and gave examples of a number of their achievements in these areas (Gordon 2010, 8–37).

Further efforts to improve scientific understanding related to climate engineering are under way, but reports from DOE, the National Aeronautics and Space Administration (NASA), National Institute of Standards and Technology (NIST), National Oceanic and Atmospheric Administration (NOAA), peer-reviewed scientific publications, and interviews with scientists indicate that the science is characterized by significant uncertainties. These gaps are related to the measurement of climate variables and models of the climate system that can simulate the effects of climate engineering on outcomes such as temperature or precipitation. The reports we reviewed described key limitations related to climate engineering science and three key challenges to improving them: (1) resolving uncertainties in scientific knowledge; (2) improving the coverage, continuity, and accuracy of observational networks used to

measure essential climate mechanisms; and (3) developing greater high-performance computational resources and dedicating them to climate modeling.

3.3.1 Better models would help in evaluating climate engineering proposals

Best practices in technology development recommend thoroughly testing new technologies before employing them in essential systems (GAO 1999). Tests usually involve controlled experiments to understand how a technology being developed works and to assess its performance. However, large-scale field testing of climate engineering technologies is difficult (Gordon 2010, 3-4, 20, 27, and 32). For example, according to NIST scientists we interviewed, estimations of or assumptions about relevant chemical, physical, and optical properties that are acceptable for many common applications would introduce unacceptable risk in large-scale climate engineering experiments that could permanently alter the chemistry of the atmosphere.

Complex climate models such as general circulation models (GCM) can be used to simulate the effects of large-scale climate engineering proposals and evaluate them without deploying them. However, the models are only as good as the data and the scientists' understanding of how the climate system works (Meehl and Hibbard 2007; GAO 1995). Scientists attending the Aspen Global Change Institute's 2006 session on Earth System Models said that gaps in climate models or inadequate data could affect the outcomes of numerical simulations designed to test climate engineering proposals (Meehl and Hibbard 2007).

General circulation models of Earth's climate evolved from short-term weather forecasting models first developed almost half a century ago (Slingo et al. 2009; McGuffie and Henderson-Sellers 2001). Advances in computing power and scientists' understanding of the climate system have helped improve the models' simulation capabilities (Slingo et al. 2009), but according to a NOAA official these improvements are still not sophisticated enough to rely on for climate engineering. Atmosphere-ocean general circulation models (AOGCM) are today's standard in climate models; they typically account for a number of factors that can influence the climate, such as oceans, land surface, and sea ice (Bader et al. 2008; Meehl and Hibbard 2007). Since 2000, AOGCM simulations have included aerosol effects, terrestrial processes, ocean mixing, and sea ice movement, but reports show that these models have important limitations with implications for simulations of the effects of climate engineering technologies.

For example, simulations of aerosol-based SRM technologies require not only a thorough understanding of how aerosols behave in the atmosphere but also a computationally intensive representation of this behavior in a climate model. At present, aerosol treatment is not standardized across GCMs, and the models generate different results in terms of predicted temperature changes and precipitation patterns (Kravitz et al. 2011). Climate engineering researchers are beginning to standardize modeling scenarios that describe actions to manipulate the climate. This standardization would allow researchers to compare the robustness of the models' responses to engineered inputs and to investigate how simplifying assumptions and structures used in the models can influence these outcomes (Kravitz et al. 2011). One scientist noted that climate chemistry models focusing

on atmospheric processes can also contribute to scientific understanding of aerosols but can be computationally intensive.

Earth systems models (ESM) representing the forefront in climate models aim to account for biological and chemical processes, such as the carbon cycle, that are not typically present in AOGCMs (Bader et al. 2008; Meehl and Hibbard 2007; Washington 2006). Climate models that included these additional processes could help scientists discover consequences of climate engineering proposals that are not predicted by the current generation of models (Meehl and Hibbard 2007). For example, simulations of CDR-based proposals could be influenced by improving the representation of the carbon cycle in climate models (Bader et al. 2008).

Scientists have identified several potential advancements related to ESMs that could improve their use in evaluating climate engineering proposals:

- scientific knowledge that would facilitate improvements in computational algorithms that represent physical, chemical, or biological processes;

- improvements to observational networks that measure essential climate mechanisms;[47] and

- greater high-performance computing resources dedicated to climate engineering-related science.

---

[47] Gaps or deficiencies in observational networks could also interfere with the ability to monitor the effect of deployed climate engineering technologies. Monitoring would allow scientists to verify the effectiveness of technologies and help ensure their safety.

3.3.2 Key advancements in scientific knowledge could help improve climate models

Although scientific knowledge of Earth's physical, chemical, and biological processes has increased over time, it remains characterized by substantial gaps that can affect measures of climate sensitivity simulated by climate models (NRC 2010a; Bader et al. 2008; Solomon et al. 2007; Meehl and Hibbard 2007).[48] Increased scientific knowledge about a number of environmental processes could improve scientific confidence in estimates of climate sensitivity. For example, Bader and colleagues (2008) highlighted the importance of improving representations of terrestrial, oceanic, and atmospheric carbon-feedback processes for more reliable estimates of future climate change. About half of all anthropogenic carbon emissions are sequestered in terrestrial or oceanic sinks whose mechanisms and capacities are not adequately revealed by observations (NRC 2007). Similarly, the relative magnitude of Earth's energy reservoirs and the exchanges between them are not fully understood (Trenberth and Fasullo 2010).[49] Scientists' limited understanding of how aerosols and clouds affect Earth's energy budget and hydrological cycle is the most important source of uncertainty in climate models (NRC 2007). Aerosols may affect climate to the same degree as CO_2 at current levels, but uncertainty about the effect of aerosols is about five times greater than the corresponding uncertainty about CO_2 (NRC 2007). Experts at a NASA workshop reported that using climate models to simulate

and evaluate aerosol-based SRM proposals to modify the climate is limited by the lack of models that explore how these aerosols would affect stratospheric ozone and the biosphere (Lane et al. 2007).

3.3.3 Better observational networks could help resolve uncertainties in climate engineering science

Observational sensing systems such as satellites and ground-based stations collect data that help scientists track climate trends and model climate mechanisms. Scientists have expressed several concerns about the coverage, continuity, and accuracy of observational networks that gather data related to climate mechanisms that are central to climate engineering technologies.

Observational network abilities depend in part on where sensors are placed and the density of their distribution (OSTP 2010; Ohring 2007). Climate engineering scientists have expressed concern the adequacy of observational networks in the atmosphere (Gordon 2010, 23). For example, some scientists have criticized the sparse distribution and output of sensors in the upper atmosphere, where a number of processes have implications for CDR and SRM technologies (NRC 2010a). In particular, scientists from NOAA and Oak Ridge National Laboratory said that CO_2 measurements from these sensors may be insufficient to permit conclusive statements about the effects of a given CDR technology. Upper atmosphere observations of the types of aerosols under consideration in some SRM proposals are also rare. Moreover, instruments that measure the optical properties of aerosols were recently eliminated from two satellites in the Joint Polar Satellite System (Gordon 2010, 15).

---

[48] One example of a measure of climate sensitivity would be "the response of global mean temperature to a doubling of [the atmospheric concentration of] carbon dioxide" (Bader et al. 2008, 2).

[49] A discrepancy in carbon output and uptake by Earth's systems remains unresolved. To preserve mass balance in today's best estimates of the global carbon budget requires including an unknown terrestrial carbon sink of about 1.8 billion tons of carbon per year (R. A. Houghton et al. 1998).

Scientists have also expressed concern about the continuity of measurements by observational networks. Scientists have noted that deferring the implementation of adequate observational networks could miss opportunities to collect data on infrequent and unpredictable natural events, such as large volcanic eruptions, that could help scientists understand mechanisms related to climate engineering (Asilomar Scientific Organizing Committee 2010; Gordon 2010, 23).[50] Scientists have also criticized the lack of redundancy in observational networks, which could create a gap in the measurement record if a single satellite or sensor were to fail (OSTP 2010; Ohring 2007; NRC 2007). For example, NASA's Glory Climate Satellite, intended to collect data on aerosols and solar energy in the atmosphere, recently failed to reach orbit at its launch.

The continuity of measurements can also be affected if programs to collect data are not sustained over a long period of time. For example, federal budget cuts in the past decade have cancelled, delayed, or degraded the collection of data from NASA's Earth Observing System satellites, whose instruments and sensors measure essential climate variables such as Earth's radiation budget, the global distribution of CO_2, concentrations of methane and other greenhouse gases, air temperature and moisture content, cloud cover, and sea surface temperatures (NRC 2010a). Further, a 2007 NRC report predicted that the nation's system of environmental satellites could decline dramatically and the number of operating sensors and instruments on NASA's spacecraft would decrease by about

40 percent by 2010 (NRC 2007). Scientists have also noted the difficulty of comparing continuous observations measured by different satellites or sensors without any overlap in their observation periods.

Scientists have expressed concern about the accuracy of data collected from existing sensing devices (NRC 2007). For example, according to NIST scientists, unknown drifts in instrument data can cause measurements to show misleading evidence of change or false trends. Additionally, because most operational or weather satellite-based sensors share a common heritage, an artificial trend in a reading from one sensor is likely to exist in similar readings from other versions of the sensor, which would bias the measurements if the drift remained undetected. Moreover, large variations exist in solar radiation measurements even over small geographic areas and the causes are uncertain.[51]

Satellite programs developed to monitor and track local weather patterns might not be accurate or precise enough to measure long-term global climate change (Fraser et al. 2008). Climate-relevant signals are extremely small compared to fluctuations in weather and temperature observed daily, seasonally, or annually (Ohring 2007). For example, a decade's anticipated average global temperature change is about 0.2 degrees Celsius, or about 1/50th of the temperature change that accompanies typical weather events. It is similarly difficult to accurately measure small variations in incoming or outgoing solar radiation on the order of 0.01 percent over decades without

---

[50] The eruption of Mount Pinatubo in 1991 released a large quantity of sulfate aerosols into the atmosphere, causing average global temperatures to fall. Scientists attending the 2010 Asilomar conference said that current observational networks are inadequate to collect data following such an eruption that could help improve scientific knowledge about atmospheric mechanisms related to aerosol-based SRM technologies.

[51] According to NIST scientists, both ground- and space-based measurements exhibit these types of variation. The space-based variations are largely attributable to calibration inaccuracies that can largely be corrected by adjustments using measurements taken during satellite overlap. The ground-based variations are both geographic and temporal and are likely to include contributions from global dimming, urban aerosols, and sensor calibration inaccuracies.

adequate optical instruments. Measuring radiation with accurate sensors is critical to advancing climate science: IPCC has reported that most climate change uncertainty derives from changes in Earth's outgoing broadband radiation (J. T. Houghton et al. 2001).

Within the federal government, steps are being taken to resolve some of these concerns, which might improve the ability to assess climate engineering technologies or proposals. Various agencies have proposed a long-term measurement strategy in Achieving Satellite Instrument Calibration for Climate Change (Ohring 2007). Satellite missions are being designed to help calibrate and reconcile some of the data received from existing climate measuring devices. For example, NASA's Climate Absolute Radiance and Refractivity Observatory (CLARREO) mission is intended to yield a benchmark data record for detecting, projecting, and attributing change in the climate system.[52]

CLARREO would constitute a major effort to correct systematic biases and discontinuities in satellite-based climate measurements and to provide a robust climate reference point for future sensors that is traceable to accepted physics-based standards, called the International System of Units. Traceability ensures that environmental measurements are comparable, independent of the organization or country making them. The National Science Foundation (NSF) is sponsoring the construction of an integrated, Earth-based observation system called the National Ecological Observatory Network that will collect data across the United States on climate and land use changes and the effect of invasive species on

natural resources and biodiversity.[53] According to NSF, it will be the first observatory network that can both detect and forecast ecological change on a continental scale over multiple decades. Gordon notes that the network could inform research on several climate engineering technologies, including land-use management and biochar (Gordon 2010, 10).

3.3.4 High-performance computing resources could help advance climate engineering science

Advances in computing could help scientists improve models used to simulate essential climate mechanisms and outcomes related to climate engineering. Limits in computational resources demand that existing climate models simplify certain processes essential to climate engineering instead of computing them numerically (Bader et al. 2008). Unlike short-term weather modelers, climate modelers have not moved to higher resolutions.[54] Instead, they have used modern computational power to include additional physical components in the calculations.[55] This is particularly important for climate engineering where, for example, stratospheric chemistry (to treat stratospheric aerosol processes) and the hydrological cycle (to treat cloud brightening) are important. A typical climate model represents Earth's system as a grid of boxes anywhere from 100 to 300 kilometers on a side, which is larger

---

[52] The President's fiscal year 2012 budget request for NASA cut much of the funding for the CLARREO mission and called for an extended preformulation period for the mission and science team to identify implementation options for obtaining climate change measurements without using CLARREO satellites.

[53] The network is expected to be fully operational in 2016.

[54] Grids for atmospheric circulation models have been refined from resolving areas the size of Colorado in 1990 to the size of South Carolina in 1995, and to about the size of Rhode Island (4,000 km²) in 2007. Meanwhile, weather models have been run with a resolution of less than 1,000 km² for over 20 years.

[55] More computer resources can be used for finer numerical grids, greater number of runs for statistical estimation, or more climate processes; consensus on the optimal resource allocation does not exist.

than a typical cumulus cloud of about 1 square kilometer (Slingo et al. 2009; Bader et al. 2008). Other significant climate features like oceanic eddies also act on a much smaller scale than the resolution current climate models support. Finer resolution that could be supported by increased computing power would help improve climate models' representations of atmospheric and oceanic circulation (Bader et al. 2008).

Computing advances could also facilitate the use of climate models to predict outcomes for geographic regions or across shorter time intervals. Scientists at NOAA's Earth Systems Research Laboratory said that finer resolution could improve climate models' predictions of regional changes that could be useful in evaluating climate engineering proposals. A climate scientist and a systems engineer also noted the potential value for climate engineering of greater precision in predicting regional changes in temperature and hydrological processes than existing models provide. They also observed that studies of climate engineering could benefit from simulations over shorter time intervals than are used in existing models.[56] Officials at NOAA have predicted that at the historical rate of increase in computing power, supercomputers able to run cloud-resolving ESMs with a grid size of a few kilometers should be available by 2025.

Some federal agencies are already developing tools that would take advantage of anticipated, massively parallel, fine-grain computational architectures based on thousands of graphics

processing units (GPU).[57] For example, NOAA is encoding an ESM to operate with small amounts of local memory that will allow it to run on these GPUs. Additionally, DOE, NOAA, and the National Center for Atmospheric Research purchased supercomputers such as the Cray XT6 and Cray Baker computers with funding from the American Recovery and Reinvestment Act of 2009, and scientists at these agencies hope to develop an ESM by 2025.

As table 3.1 and table 3.2 show, climate engineering technologies are in early stages of development and have variable and uncertain cost factors, while uncertainty surrounds their potential effectiveness and potential consequences. Moreover, gaps in scientific knowledge, data, and computing resources challenge the models of climate mechanisms related to climate engineering.

---

[56] Existing models were designed to distinguish long-term climate trends (Fraser et al. 2008).

[57] The advent of GPUs follows on three revolutions in computing operations: (1) integrated circuits, (2) vector computing, and (3) parallel computing (which made current forecasting possible). GPUs complete intense, split-second calculations efficiently to render virtual representations of the real world.

4 Experts' views of the future of climate engineering research

Because climate engineering technologies are currently immature, we explored future prospects for climate engineering research by obtaining a wide range of expert views. We obtained expert views on the future of research in three stages: (1) 6 experts met with our Chief Scientist to construct alternative future scenarios for climate engineering research, (2) 28 additional experts representing a wide range of professional disciplines and organizational affiliations shared their views about the future in response to the scenarios, and (3) some of the experts participating in the Meeting on Climate Engineering, which we convened with the help of NAS, volunteered their thoughts about the future.[58] (In the appendices to this report, sections 8.3 and 8.4 present the scenarios and list the experts who served as scenario-builders, section 8.5 lists the experts who provided comments in response to the scenarios, and section 8.6 lists the experts who participated in the meeting we convened with the assistance of NAS.) Altogether, 45 experts contributed views on the prospects for climate engineering research across the next 20 years.[59]

Briefly, we found the following:

• The majority of the experts we consulted advocated starting significant climate engineering research now or in the very near future.[60] Among the reasons they gave for starting research now is the anticipation that two decades or more of research will be needed to make substantial progress toward developing and evaluating climate engineering technologies with the potential to reduce emerging or future risks from climate change. Research advocates also envisioned safeguards to protect against potential adverse consequences or risks arising from the research.

• A small number of those we consulted opposed starting such research, in part to prevent negative consequences either from the research or from deploying technologies developed from it.[61]

• The majority envisioned a federal effort that would direct and support research on climate engineering with specific features such as (1) an international focus, (2) engagement of both the public and decision-makers, and (3) foresight considerations to help anticipate emerging research developments and their opportunities and risks.[62]

---

[58] Among the experts we consulted, primary areas of expertise spanned two broad categories: (1) physical science or technical research related to climate engineering or climate change and (2) social science, law, ethics, or other related fields with applications in climate engineering or climate change.

[59] Although we attempted to consult diverse experts representing the full range of views on climate engineering, the relative numbers who expressed a particular view to us may not reflect the entire community of those with similar kinds of expertise. However, for transparency, we provide the specific numbers of experts who told us that they advocated certain views. We note that not all experts expressed an opinion on all issues.

[60] Two-thirds of the experts we consulted about the future (31 of 45) advocated starting significant research now or in the very near future.

[61] Four of the 45 experts we consulted about the future stated that they opposed research on climate engineering. (One of these 4 made exceptions for certain kinds of research, such as computer modeling.)

[62] Twenty-nine of 45 experts envisioned a federal research effort, and as detailed below, 26 of these mentioned one or more of these three features.

- Experts identified many trends as potentially affecting research, including the pace of climate change, emissions-reduction developments, and scientific breakthroughs.

4.1 A majority of experts called for research now

The majority of the experts we consulted about the future advocated starting significant research now or in the very near future, largely from concern about future climate change and the need to reduce its risks.[63] In this report, we define "starting significant climate engineering research" as increasing research beyond that now being conducted. We had reported earlier that a relatively small amount of federal research is directly focused on climate engineering (GAO 2010a, 19).[64] The advocates of research now—and some experts who did not indicate whether they advocated starting research now—anticipated that research will produce technologies or evaluative information or both that might help reduce risks associated with climate change or uninformed responses to it. Risks from climate change might include, for example, potential breakdowns in food and water supply chains (as climate change brings precipitation changes and rises in sea level), mass migration, and international conflict. More than half of these proponents anticipated

that substantial research progress will take time—perhaps two decades or more—or stated that we cannot and should not wait for a crisis. Additionally, research advocates indicated that a cautious risk management approach could help reduce research-related risks.

Those who advocated research now either did so urgently because they anticipated a definite need for climate engineering or viewed research as an insurance policy. For example, some warned against (1) losing the ability to prevent what they perceived as potentially irreversible changes or (2) being unprepared for a crisis. With respect to the latter, the report's scenarios (1) describe how leaders who are unprepared while under heightened pressure to act quickly in a crisis might decide to deploy inadequately understood, risky technologies and (2) present the view that informed leaders might decide not to deploy risky technologies.[65]

Others who called for research now recognized the uncertainty of the future and viewed climate engineering research

"as an insurance policy against the worst case scenarios"

in the longer-term future. One said that the nation should

"make investments [in] . . . fundamental research . . . to be able to react quickly [if needed]. . . . [Spending to limit risk] on climate, terrorism, national defense,

---

[63] Thirty-one of 45 experts said they advocated research now or in the near future, and 4 opposed research. The remainder did not clearly state whether they advocated starting research now.

[64] Specifically, we reported that 13 federal agencies had identified at least 52 research activities, totaling about $100.9 million, as relevant to climate engineering in fiscal years 2009 and 2010 (GAO 2010a)—$1.9 million for activities to investigate specific climate engineering approaches and $99 million related to conventional mitigation strategies or basic science that could be applied to improving understanding of climate engineering.

[65] We note, however, that national leaders might not base decisions on information about research results, even if such information were available.

nonproliferation, should be viewed identically."[66]

Overall, those calling for research now reflected foresight literature that warns against falling behind a potentially damaging trend, as illustrated in figure 4.1. Those advocating research now recognized important cautions, discussing two types of risk associated with research: (1) risks from conducting certain kinds of research (for example, large-scale field trials of potentially risky technologies) and (2) risks from using or misusing research results (for example, deploying risky technologies developed

from theresearch).[67] Various research advocates therefore suggested potentially complementary remedies such as

- managing risks (from research and using its results) with strategies like those outlined in box 4.1, which have been highlighted in the literature;

- evaluating the risk of deploying specific technologies, in advance, which could lead to taking some risky technologies off the table; and

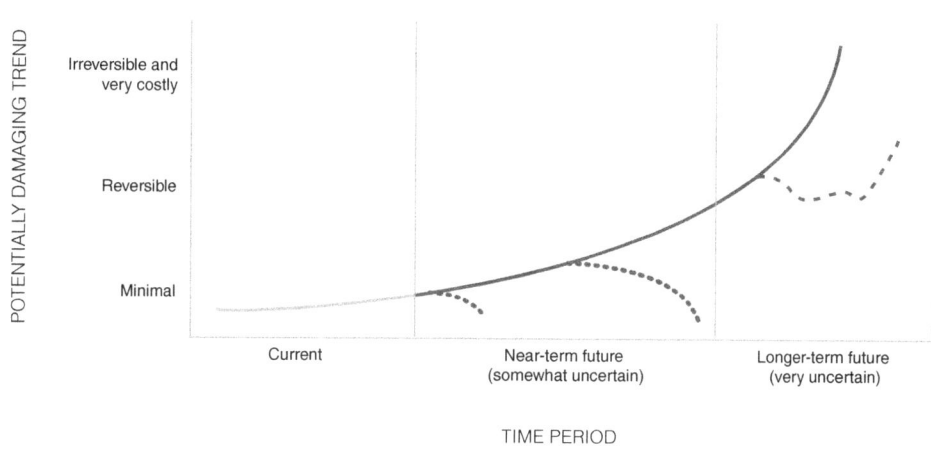

Figure 4.1 Taking early action to avoid potentially damaging trends: Illustration from foresight literature. Source: GAO adapted from Rejeski (2003).

[66] This insurance view reflects the hedging strategy described in foresight literature whereby, faced with uncertainty, decision-makers choose a strategy that they anticipate will work reasonably well across all alternatives to avoid potentially disastrous low-probability outcomes (Popper et al. 2005).

[67] Overall, of the 31 experts advocating research now, 27 recognized risks associated with it, including risks from conducting it (11 experts) and from using its results (26 experts). One advocate who believes that it is urgent to start research now also said that guidelines are needed to decide when research "has become too dangerous to continue."

```
┌─────────────────────────────────────────────────────────────────────────────┐
│                                                                               │
│   Box 4.1: Climate engineering research: Risk mitigation strategies from the literature │
│                                                                               │
│   • The research community's voluntary self-governance and development of     │
│      norms and best practice guidelines for open and safe research            │
│                                                                               │
│   • Required examinations of ethical, legal, and social implications in federally │
│      funded research projects                                                 │
│                                                                               │
│   • Interventions that bring social scientists, ethicists, or trained risk assessors directly │
│      into laboratories to ensure early accounting for risks and social and ethical issues │
│                                                                               │
│   • Application of an institutional review board concept to climate engineering research[a] │
│                                                                               │
│   • Commissioned and independently conducted interdisciplinary risk assessments │
│                                                                               │
│   • A multistage approach in which initial research (for example, computer modeling and │
│      laboratory studies) investigates risks before progressing to small-scale field studies, │
│      which in turn provide added information on risks before progressing to large-scale │
│      field experiments                                                        │
│                                                                               │
│   • Study of risk trade-offs and analysis of options for reducing overall risks │
│                                                                               │
│   • Developing norms for deployment decisions, facilitated by research activity │
│                                                                               │
└─────────────────────────────────────────────────────────────────────────────┘
```

Source: Olson forthcoming; Morgan and Ricke 2010; Victor 2008; Graham and Wiener 1995.

[a]Institutional review boards typically review research projects that use humans so as to protect their rights and welfare. However, the concept could be expanded to require an institutional review of research for field experiments that use Earth as a subject. IRB review requirements could be linked directly to federal research grants given for climate engineering.

• setting international research limitations or guidelines.[68]

As we discussed above, research advocates suggested the study of climate engineering risks, and we earlier reported (GAO 2010a) that experts had told us that potentially "unintended consequences . . . require further study."[69]

Some research advocates qualified their positive view of climate engineering research with the proviso that emissions reduction efforts be continued. They warned that if the concentration of CO_2 continues to rise into the long-term future, deploying SRM—and increasing it over time to maintain acceptable

[68] We reported earlier (GAO 2010b, 13) that in 2008 the parties to the London Convention and London Protocol issued a decision stating that ocean fertilization that is not legitimate scientific research is contrary to the aims of the agreements and should not be allowed. The treaties' scientific bodies are developing an assessment framework for countries to use and evaluate whether research proposals are legitimate scientific research (GAO 2010a, 33). In 2010, the parties to the Convention on Biological Diversity invited countries to consider the following guidance: (1) ensure that ocean fertilization activities are consistent with the London Convention and Protocol and decisions issued by the conference of the parties to those treaties and (2) ensure that except for certain small-scale scientific research studies, no climate-related geoengineering activity that may affect biodiversity take place until there is an adequate scientific basis on which to justify it and appropriate consideration of the associated risks and impacts.

[69] Additionally, of the 52 research activities federal agencies identified as relevant to geoengineering in our 2010 report, only one project's activity description specifically mentioned risk (GAO 2010a).

temperatures—would lead to serious risk. That is, should extensive SRM deployments fail or be discontinued for any reason,

"the bounce-back effect [of sudden warming] would be staggering."

Some research advocates also indicated that reducing emissions (and using apparently safer CDR technologies, such as direct air capture and sequestration, if needed to reduce build-up) would logically reduce the need for potentially risky SRM deployment.

Finally, a research advocate who reviewed this report emphasized the need to consider the net effect of (1) the risks of climate change without climate engineering, (2) the potential reduction of climate risks through climate engineering, and (3) the introduction of possible new risks through climate engineering.

4.2 Some experts opposed starting research

We noted above that a small number of the experts we consulted opposed starting significant research on climate engineering.[70] Some thought that pursuing further research would, in the words of one commenter, open a Pandora's box better left unopened. Opponents of research viewed climate engineering as technological hubris or as likely to be ineffective or not needed. Those who said that climate engineering is not needed either believed that climate change has been exaggerated or preferred other approaches, such as "building ecosystem and

community resilience to respond to climate change," "adopting more sustainable agricultural policies," or "making [a] massive investment in energy efficiency." However, the most strongly expressed opposition to climate engineering research concerned risks. One research opponent envisioned situations in which stratospheric aerosols would produce conflicts or catastrophic results in some parts of the world:

"There are wars waged over the position and density of the clouds, rainfall patterns, ocean alkalinity, and volcanic eruptions as confusion prevails over what phenomena are natural and which are manmade. Different, often conflicting experiments are sponsored by different countries . . . .

". . . the precipitation patterns over large parts of Africa and Asia, which are already suffering from drought and food insecurity . . . [are disturbed by the SRM deployment and eventually] hundreds of millions of people die because of crop failures and chaotic weather events . . . . [A] very small number of people control the climate levers, [and . . .] global tensions rise." [71]

The research opponents in our study did not envision varied strategies for managing, reducing, or avoiding risks from research or technologies

---

[70] Four of the experts we consulted opposed conducting research. Additionally, because of questions some reviewers of a draft of this report raised, we note that three of the four opponents of research had primary expertise in fields such as social science, law, ethics or other related fields (rather than physical science); these three provided the direct statements of research opponents that we quote in this section.

[71] These risks (international conflict, drought, and famine) that an expert cited as potentially deriving from climate engineering research and deployment are similar to those associated with climate change. Research opponents also pointed to other possible risks; for example, some potential SRM technologies have been associated with the depletion of ozone and interference with the use of solar-energy technology.

developed from it. One opponent of research endorsed international moratoriums and said that

> "It is illogical to assert that the best risk-avoidance strategy is to increase research. The best way to avoid responses that are extremely high risk is not to research them more; it is to make sure, through legally binding agreements, that they are prohibited."

While research advocates suggested evaluating technologies in advance of deployment, some opponents thought that, as one said,

> "the effects of human intervention are impossible to predict with a high degree of certainty. Any large-scale attempt to tame the climate system . . . has a high probability of backfiring."

Finally, some research opponents were concerned about moral hazard—that is, the possibility that the results of climate engineering research would

> "undermine the political will to reduce emissions."

Some research opponents feared that climate engineering would be substituted for, rather than used to complement, emissions reduction efforts. One research opponent suggested that

> "Political leaders . . . faced with the choice of politically difficult unilateral reductions in carbon emissions and the illusion of a techno-fix, [will] go for the latter."

As we discussed in the previous section, some advocates warned of negative outcomes if climate engineering, particularly SRM, were pursued in the absence of emissions reduction.

4.3 A majority of experts envisioned federal research with specific features

We reported earlier that the United States does not have a "coordinated federal strategy for geoengineering, including guidance on how to define . . . geoengineering activities or efforts to identify and track . . . funding related to geoengineering" (GAO 2010a, 23). In that report, we recommended that

> "the appropriate entities within the Executive Office of the President (EOP), such as the Office of Science and Technology Policy (OSTP), in consultation with relevant federal agencies, develop a clear, defined, and coordinated approach to geoengineering research in the context of a federal strategy to address climate change that (1) defines geoengineering for federal agencies; (2) leverages existing resources by having federal agencies collect information and coordinate federal research related to geoengineering in a transparent manner; and if the administration decides to establish a formal geoengineering research program, (3) sets clear research priorities to inform decision-making and future governance efforts." (GAO 2010a, 39)

OSTP neither agreed nor disagreed with our recommendation but provided technical and other comments. With respect to the context of a federal strategy to address climate change, we note that other approaches to addressing climate change include efforts to (1) reduce CO_2 emissions and (2) adapt to climate change.

In our work for this report, we found that experts who advocated starting significant research now generally also advocate or envision a federal research effort with specific features.[72] That is, they envision federal research that would foster the development of technologies like CDR and SRM, rigorously evaluate related risks, and include specific features such as

- an international focus,

- engagement of the public and national decision-makers, and

- incorporation of foresight considerations aimed at identifying new opportunities, anticipating new risks, and adapting research to emerging trends and developments.

As outlined in this report's overview of the technologies, knowledge is currently limited on proposed CDR and SRM technologies, and experts' comments to us further suggested that planning might need to precede the first phase of any federal research effort.

Some research advocates who envision a U.S. climate engineering research effort explained that other nations, the United Kingdom among them, are already studying these technologies or establishing programs.[73] They said that the absence of a U.S. research effort could leave

the United States "without a seat at the table," unprepared to play a leading role, or unable to respond to other nations' actions. One advocate of a U.S. research effort said that

> "If it ever becomes necessary to deploy geoengineering techniques, doing so will be a momentous decision for humanity. The United States should be prepared to play a leading role in the decision, and it should be unthinkable that the decision could be made without substantial input from the U.S. scientific and technical community."

Others imagined a future in which individual nations would unilaterally engage in SRM; one motivation might be to resolve local or regional problems caused by climate change (Morgan and Ricke 2010). On one hand, such actions could have transboundary or global SRM effects, conceivably raising issues of national security, stimulating other nations to respond, or requiring a U.S. response. On the other hand, the risk of unilateral action might be reduced with cooperative international research that fostered trust and cooperation among nations on issues pertaining to climate engineering.[74]

For reasons such as these, many of the research advocates in our study suggested an international approach to federally sponsored research.[75] Suggested activities included

---

[72] Of the 31 experts who advocated starting research now, 29 also advocated or envisioned a federal research effort; 26 of these envisioned one or more of the three specific features discussed in this section. Some experts also anticipated the development of technologies by the private sector.

[73] One expert told us that some nations' research may be hidden because it is not specifically labeled as climate engineering or because it is covert.

[74] One approach to international research cooperation is illustrated by the International Space Station, with its five main partners: Canada, Japan, Russia, the United States, and the European Space Agency (which includes a number of countries) (GAO 2009c).

[75] Twenty-four experts (of 31 who advocated climate engineering research now) specifically envisioned an international approach for federal research.

- studying strategies for responding to situations arising from insufficient international cooperation in the use of climate engineering;

- sponsoring or encouraging joint research with other nations (including developing or emerging industrial nations) because this might (1) help the United States keep pace with other nations' research; (2) facilitate rigorous, transparent evaluation of new technologies that others develop; and (3) foster cooperation and consensus—or an evolving set of norms about conducting research, which might, in turn, foster support for guidelines;[76] and

- studying how the responsibilities of nations that deploy these technologies could impinge on others' geopolitical equity, human rights, and justice—which would logically be most important for vulnerable or poor populations.

Other international issues suggested for research included (1) studying how to define climate emergencies and achieve international agreement on responses to them and (2) exploring issues concerning military engagement in climate engineering research.

The possibility of U.S. leadership in internationally focused research was suggested, as was cooperation:

"What the U.S. can do . . . is to lead the process of framing the [climate engineering] issue as one requiring global collaboration and evidence-based decision-making processes that focus not only on macro results but also on fairness in distributional aspects of action versus inaction."

One expert said that in a substantial, internationally focused U.S. research effort, the United States could lead by example, emphasizing values such as transparency and attention to risk issues.

The engagement of researchers with the public and with U.S. decision-makers (and possibly international leaders) was another desirable focus for a federal research effort, according to research advocates.[77] Their views included statements that

- engagement can foster shared learning across national leadership, the general public, and the research community; help ensure transparency; build shared norms; and bring an informed

"democratic process [to] . . . decisions that . . . broadly affect society;"

- engagement results might help frame research agendas to reflect the concerns and needs of the public and decision-makers; and

- information provided to the public and decision-makers might address

---

[76] One of our scenarios describes the lessened possibility of conflict because nations, having cooperated on research, have a basis for cooperating in a sudden crisis. Our 2010 report indicated that "several of the experts we interviewed as well as the NRC study emphasized the potential for international tension, distrust, or even conflict over geoengineering deployment" and discussed international agreements and governance challenges (GAO 2010a, 17 and 26–37).

[77] Twenty-three experts (of 31 who advocated starting research now) favored engaging the public or national leaders or both in a federal effort.

- (1) the systemic risks of the various climate engineering approaches, (2) trade-offs in pursuing alternative strategies, and (3) analyses of ethical, economic, legal, and social issues.

A broad, multidisciplinary research agenda consistent with these views is discussed in the foresight scenarios we developed for this report.

Our discussion of risks (earlier in this section) indicated that uninformed national or global leaders who were under heightened pressure in a perceived climate crisis might make hasty choices, whereas informed leaders might make a more measured response. Logically, the same might be true of the general public. That is, public engagement in advance of a crisis could help ensure that public concern about harm from technologies is addressed in advance (through research on benefits and risks), that research results are appropriately conveyed to the public, and that public expectations are consistent with likely real-world consequences. In sum, communication among researchers, the public, and decision-makers might help prepare the nation for a measured response to a future crisis.

Some advocating a federal research effort also envisioned its incorporation of foresight activities designed to (1) anticipate emerging directions in developing climate engineering technologies and new or changing risks associated with such directions and (2) help research keep pace with other developing trends that could affect the research agenda and support.[78] Some examples of foresight activities are communicating with other researchers in related areas, monitoring or surveying research, and using horizon scans and other futures methods that could help anticipate and track relevant developments and potential new risks.[79] One research advocate also suggested exploring low-probability, high-impact events (described by Taleb 2007) with game theory or scenario planning.

Other examples include iteratively monitoring a variety of developments and trends and, where appropriate, supporting studies that obtain better evidence on them. This could help guide decisions about forward directions (GAO 2008b, 67–68) and is compatible with other suggestions for adaptively managing climate engineering research. (With respect to the latter, experts in our study endorsed adaptive management to better achieve continuous improvement, based on (1) changing practices over time in response to experience and performance assessment and (2) learning how to intervene in a complex, imperfectly understood climate system.[80])

Finally, the overall results of our communications with experts indicated uncertainty, or at least a diversity of views, on what technical research and evaluation will be needed for specific CDR and SRM technologies. Some experts noted the lack of any map or forward-looking plan showing how climate engineering research might progress along various paths. We also found that experts expressed widely different views on

---

[78] Eighteen experts (of 31 who advocated starting research now) specifically envisioned an anticipatory, foresight approach for federal research.

[79] A horizon scan is a systematic examination of ongoing trends, emerging developments, persistent problems that may have changed, and novel and unexpected issues. Horizon scans are sometimes structured strategically to consider potential threats and opportunities separately.

[80] For example, Long (2010) has said that an adaptive approach is appropriate for climate engineering because climate is a complex, nonlinear system. Such an approach might include monitoring the results of an intervention, comparing observations to predictions, deciding whether the research is proceeding in the right direction, and making a new set of decisions about what to do.

- the scope of whatever global climate engineering efforts might eventually be implemented or deployed and

- the level of effort or funding needed for research.[81]

For example, experts variously characterized the scope and scale of the deployment of stratospheric aerosols in terms of (1) operations that "rogue" actors might carry out unilaterally or, in contrast, (2) huge operations that might amount to "the largest engineering project in the history of people."

Some experts we consulted suggested that research funding might start with as little as a few million dollars. The scenarios developed for this report suggest that more effective research would have a considerably higher budget but they do not specify an amount. We reported earlier that 13 federal agencies had identified at least 52 research activities relevant to climate engineering in fiscal years 2009 and 2010 (GAO 2010a)— with funding of $1.9 million to investigate specific climate engineering approaches. Much larger amounts have funded activities related to conventional mitigation strategies or basic science that could be applied to improving scientific understanding of climate engineering.[82]

Because information about climate engineering and related research is limited, one of the experts who advocated federal research suggested that a federal effort begin with initial developmental work to delineate scale and cost. (Further research might then be planned and potential research

costs estimated (GAO 2009b).) According to another expert, planning efforts would benefit from the development of an overall research strategy, including, for example, a

"multidisciplinary framework for integrated systems analysis . . . and risk assessment tailored to designing and evaluating geoengineering technologies and their potential deployment as subscale experiments."

4.4 Some experts thought that uncertain trends might affect future research

When the experts we consulted envisioned research with a foresight component, they saw the following as relevant and potentially critical to track. First, signals of impending climate-related events would be relevant because these could potentially heighten the urgency and priority of the research. An example might be a collapse of ocean fisheries attributed to global warming and ocean acidification, with depletion of food supplies in vulnerable areas. Second, trends in policies for or new approaches to emissions reduction could affect prospects for CDR's implementation. Our scenarios illustrate the view that establishing carbon constraints would encourage an anticipation of the use of CDR research, creating an incentive for research and innovation.[83] Experts differed in assessing how CDR research might develop in the absence of significant carbon constraints. Some said it

---

[81] Additionally, 10 experts told us that either analytical information on the cost of a potential climate-engineering research program is lacking or they did not know of such information.

[82] An additional $99 million supported these other activities.

[83] Carbon constraint policies aim to limit or reduce carbon emissions. Greenhouse gas emissions pricing is one type of carbon constraint that would encourage people to reduce emissions by making them more expensive. Despite ongoing debate over climate change legislation, the U.S. Congress did not enact legislation in 2010, and its prospects are uncertain.

would be difficult to sustain research or deploy CDR technology without carbon constraints, while others disagreed, citing the possibility of deployment through a major public works program (Parson 2006). Also relevant would be developments in sequestration related to advances in carbon capture and storage.

Other potentially important areas to track are nanotechnology and synthetic biology breakthroughs (Rejeski 2010; Shetty et al. 2008); advances in these areas might bring new developments in climate engineering technologies. Examples include future "programmable plants" that would sequester more carbon than natural plants and airborne microbes that would consume greenhouse gases. However, such developments might entail new risks.

Future research breakthroughs might lead to or create new low-cost, low-carbon technologies and thus speed emissions reduction. One expert envisioned no-carbon energy sources like solar power costing less than carbon-based energy. Developments such as these could have important implications for the future role of CDR.

Additionally, experts thought trends in public opinion on climate engineering research might affect support for research or specific projects. Monitoring trends in public opinion could be a key element of public engagement; for

example, it might signal a need to study the safety implications of certain kinds of studies.

Finally, experts (1) suggested links between future developments in climate engineering and possible international tensions or conflicts that might develop from economic issues, cultural changes, or demographic shifts and (2) indicated that low-probability, high-impact events might affect future research. They suggested examples of the latter, such as an SRM experiment's coinciding with a natural volcanic eruption and producing unprecedented cooling; abrupt changes in ocean currents sharpening climate differentials; catastrophic alterations in weather patterns; geopolitical instability caused by widespread and prolonged famine in Africa or the Indian subcontinent attributable to global warming; a biotechnology disaster's leading to strong public sentiment against technological interventions; the low-cost distribution of locally affordable technology's reducing shipping-related carbon emissions; or sudden cooling from an asteroid hit.

If research planners believe that some low-probability events represent sufficient risks or opportunities, they might decide either on contingency planning or on hedging—that is, selecting a strategy that works reasonably well across a variety of outcomes, including certain low-probability, high-impact events (Popper et al. 2005).

5 Potential responses to climate engineering research

Because climate engineering technologies are potentially risky and could affect a large number of people and because experts have noted the importance of public engagement on this issue, we collected baseline measures of public opinion on climate engineering research among U.S. adults today. We analyzed survey responses from 1,006 U.S. adults 18 years old and older (representing the U.S. public) to address our third objective concerning the extent of awareness of geoengineering among the U.S. public and how the public views potential research into and implementation of geoengineering technologies.[84]

We found that the majority of the U.S. public is not familiar with geoengineering. Because public understanding of geoengineering is not well developed and public opinion in this area may be influenced by a variety of factors that may change over time, it is important to note that the results we report are not intended to predict future U.S. public views. Rather, our results provide valuable baseline information about current awareness of geoengineering and how the U.S. public might respond if it learned more about geoengineering.

Because the public lacked familiarity with geoengineering, we provided survey respondents with basic information about geoengineering technologies before asking questions about them.

Our key findings are that if the public were given the same type of information that we gave our survey respondents, then

- about 50–70 percent of the U.S. public across a range of demographic groups would be open to research on geoengineering.[85] Many survey respondents expressed concern about the potential for harm from geoengineering technologies, but a majority also said they believe research should be done to determine whether these technologies are practical;

- about half of the U.S. public would support developing geoengineering technologies. At the same time, about 75 percent would support reducing CO_2 emissions and increasing reliance on solar and wind power;

- about 65–75 percent of the U.S. public would support a great deal, a lot, or a moderate amount of involvement by the scientific community, a coalition of national governments, individual national governments, the general public, and private foundations and

---

[84] Knowledge Networks Inc. fielded the survey of a statistically representative sample of 1,006 respondents July 19 to August 5, 2010, using its online research panel. We used the term "geoengineering" in our survey questions and other information we provided about climate engineering because we had used the term in earlier work. All estimates from the survey are subject to sampling error. In terms of the margin of error at the 95 percent confidence level, the sampling error for estimates based on the total sample is plus or minus 4 percentage points and, for estimates based on subgroups of the sample, is plus or minus 9 percentage points, unless otherwise noted. Because the overall response rate was low and sources of nonsampling error may have contributed to total survey error, we rounded survey results to the nearest 5 percentage points. We describe our methodology in more detail in section 8.1.3.

[85] The sampling errors for the following demographic subgroup estimates, in terms of the margin of error at the 95 percent confidence level, are plus or minus 12 percentage points for the percentage of those with less than a high school education who believe research should be done on geoengineering, plus or minus 13 percentage points for the percentage of blacks who believe research should be done on geoengineering, and plus or minus 14 percentage points for the percentage of Hispanics who believe research should be done on geoengineering. The margin of error for the remaining subgroup estimates is plus or minus 9 percentage points.

not-for-profit organizations in making decisions related to geoengineering.

5.1 Unfamiliarity with geoengineering

Many people in the United States believe that Earth is warming but are not certain that this can be changed, while others do not believe that global warming is happening (Leiserowitz et al. 2010, 7; Maibach et al. 2009, 1 and 13; Nisbet and Myers 2007, 451). National surveys of U.S. public opinion have found broad public support for a variety of measures to increase energy efficiency, diversify the energy supply, and reduce CO_2 emissions (Pew 2010, 3; Bittle et al. 2009, 11), but geoengineering has not yet received widespread attention.

Given the diversity of views on climate change, our survey asked respondents to consider their own views on climate change and how serious climate change might be and to indicate whether they thought any action should be taken. From the responses to this question, we estimate that about 40 percent of the U.S. public thinks that immediate action on climate change is necessary, about 35 percent thinks that action should be taken only after further research, about 10 percent thinks that no action should be taken, and about 15 percent is unsure. Among those who do not believe the climate is changing, we estimate that about 50 percent thinks no action should be taken and about 40 percent thinks that action should be taken only after further research. In other words, members of the U.S. public who do not believe the climate is

changing do not necessarily oppose research on climate change.[86]

To ensure that our survey respondents had a basic understanding of geoengineering, we gave them a brief definition of geoengineering and examples of CDR and SRM technologies before we asked them questions about geoengineering. The information we gave them was similar in amount and type to information they might receive in the nightly news or in a short news article.

Immediately after we defined geoengineering for our respondents and gave them examples of CDR and SRM technologies, the survey asked them whether they had ever heard or read anything about geoengineering technologies before they began the survey. From the results, we estimate that if provided with information about geoengineering similar to that given our survey respondents, about 65 percent of the U.S. adult public would not have recalled hearing or reading anything about geoengineering technologies at the time of our survey. The results of our survey pretest interviews, which included follow-up questions, indicated that some members of the public recall reading or hearing about technology proposals such as sequestration of carbon in the ocean or other geoengineering-type technologies in science and technology literature.

---

[86] Because our focus for this report was on public perceptions of climate engineering, our survey was not designed to assess public views of climate change more broadly. It did, however, ask several questions about climate change and energy policy similar to those in prior surveys of the U.S. adult population. While comparisons between our survey and others' surveys are not conclusive because of historical, methodological, and measurement differences, we found a general similarity between the distribution of our results and those from other sample surveys.

5.2 Concern about harm and openness to research

As identified above, climate engineering includes a number of technologies, and different technologies may have different risks and benefits. To assess whether information about the potential for harm from different technologies affects public reaction to climate engineering, we decided to conduct a split-ballot survey in which we gave half the sample information about technologies that had been identified as relatively safe and the other half information about technologies that had been identified as less safe.[87]

This allowed us to examine whether receiving information about the less safe or the more safe technologies is associated with greater concern about harm from geoengineering. It also allowed us to assess whether public opinion on research and decision making depends on the information members of the public are given about experts' assessments of a technology's relative safety.

We differentiated technologies by the experts' assessments of safety as described in the Royal Society report (Royal Society 2009, 6). The two relatively safe technologies in our survey were (1) increasing reflection from Earth's surface (by painting roofs, roads, and pavement white, for example) and (2) capturing CO_2 from the air (in the information we gave the respondents, we also called this CO_2 air capture and capturing CO_2 from the air). The two less safe technologies

were (1) putting sulfates, or tiny mirror-like particles, into the stratosphere and (2) seeding large ocean areas with fertilizer. Table 5.1 shows the information the respondents received about technology by the ballot group they were assigned to—506 respondents received information about increasing reflection from Earth's surface and CO_2 air capture, and 500 received information about stratospheric sulfates and ocean fertilization.

We randomly assigned survey respondents to receive information about the relatively safe and the less safe technologies. At the outset of receiving the information about geoengineering, survey respondents were told that

> "Some scientists believe it might be possible to deliberately change Earth's temperature and cool down the planet by changing some of the things that seem to be causing global warming. Using technologies to do this is known as 'geoengineering.'

> "There are two different types of geoengineering. The first type involves reflecting some of the light and heat of the sun's radiation back into space. The second involves reducing the level of carbon dioxide in the atmosphere."

The respondents were not aware that the survey had two different sets of examples of geoengineering. All the survey questions were identical.

---

[87] We did not vary effectiveness in the split-ballot design. In each ballot group, respondents learned about one technology that the Royal Society's 2009 report had identified as highly effective (either capturing CO_2 from the air or injecting stratospheric aerosols) and one that it had identified as relatively less effective (either increasing reflection from Earth's surface or fertilizing the oceans). The design did not allow us to determine how experts' assessments of the different technologies' effectiveness might affect public reactions to geoengineering.

Example	Technology type	
	Relatively safe (506 respondents)	Less safe (500 respondents)
Reflecting back into space some light and heat from the Sun's radiation	**Increasing reflection from the surface of Earth** Increasing reflection from the surface of Earth involves lightening and brightening the surface of the earth, to reflect some of the sunlight back into space. By reflecting sunlight into space, the temperature would be reduced. Increasing reflection from the surface of Earth could involve painting roofs, roads, and pavement. Although this should reduce the temperature at least some, there are doubts whether reflecting the surface of Earth could have a substantial effect on global temperatures. Unlike some geoengineering techniques, however, there is little risk of negative consequences. So the technique of reflecting the surface of Earth is not very effective, but it is safe.	**Putting sulfates into the stratosphere** Putting sulfates, which are tiny mirror-like particles, into the stratosphere. This would re-create what happens when large volcanoes erupt and shoot sulfates high into the atmosphere. The sulfates circulate in the stratosphere and reflect some sunlight before it reaches Earth. Research has shown that this technique would probably be very effective at reducing the global temperature. The extent and type of consequences from stratospheric sulfates is unknown, however. For example, there could be increased damage to the ozone layer or altered rainfall patterns around the world. So the technique of stratospheric sulfates is likely to be very effective, but there is also risk of serious negative consequences.
Reducing carbon dioxide in the atmosphere	**Capturing carbon dioxide from the air** CO_2 air capture would chemically remove CO_2 directly from the air. The CO_2 could be turned into a liquid and piped underground for storage in geologic structures. This technique directly treats the cause of climate change—greenhouse gases— and research has shown that CO_2 air capture would be very effective. Unlike some geoengineering techniques, it would not directly affect complex natural systems and is believed to be safe. So the technique of CO_2 air capture is likely to be very effective, and it is safe.	**Seeding large ocean areas with fertilizer** Ocean fertilization involves adding nutrients such as iron to some areas of the open ocean where they are in short supply. This promotes the growth of small plants called phytoplankton, and as the plants grow, they soak up CO_2 from the atmosphere. This technique directly treats the cause of climate change—greenhouse gas such as CO_2. It is not yet known how much carbon would be removed for longer than a few years; we need to learn more about the effectiveness of ocean fertilization. The extent and type of consequences from fertilizing the oceans are also largely unknown. For example, there may be harmful side effects if ocean fertilization were attempted on a large scale. So the technique of ocean fertilization may not be very effective and there is also the risk of serious negative consequences.

Table 5.1 Geoengineering types and examples given to survey respondents. Source: GAO.

Note: The information provided to respondents was based on a report from the Royal Society (Royal Society 2009).

The survey results indicated that some 50 percent or more of both survey ballot groups were somewhat to extremely concerned that geoengineering could be harmful. More specifically, we estimate that

- about 50 percent of the U.S. adult public would be somewhat to extremely concerned that geoengineering technologies could be harmful if they were given information similar to what we gave the respondents about relatively safe technologies (increasing reflection from Earth's surface and capturing CO_2 from the air) and

- about 75 percent of the public would be somewhat to extremely concerned that geoengineering technologies could be harmful if given information similar to what we gave respondents about less safe technologies (stratospheric sulfates and ocean fertilization).

These results suggest that many people would be concerned about the safety of even technologies that experts have identified as relatively safe. For technologies experts deemed less safe, a substantial majority would express concern.

Despite these differences in respondents' concerns, they did not differ greatly in their responses to other questions about geoengineering research and decision making. Consequently, we report the results from all other survey questions for all survey respondents combined.

In addition to the issue of the technologies' harm, the survey asked respondents how optimistic they were that geoengineering technologies could be beneficial. From the results, we estimate that about 45 percent of the public would be somewhat to extremely optimistic, about 40 percent would be slightly to not at all optimistic,

and about 15 percent would be unsure whether geoengineering technologies could be beneficial. As reflected in responses to an open-ended question in our survey, public optimism about geoengineering is likely to be tempered by concern that the technologies' effects are not fully known. As one survey respondent put it:

> "Since the outcome is uncertain, more research needs to be done to find out how much of any one thing is enough or too much."

Given that research may be seen as a way to assess whether specific technologies might work and to identify harmful consequences, we used the survey to identify a baseline estimate of support for research on geoengineering among the U.S. public. From the results, we estimate that about 65 percent of the public, exposed to the same type of information as in our survey, would say they believe that research should be done to determine whether geoengineering technologies that deliberately modify the climate are practical. Further, respondents who received information about less safe technologies were just as likely to support research to determine whether geoengineering is practical as were respondents who received information about safer technologies; moreover, about 60 percent of those who said they were extremely concerned that geoengineering could be harmful indicated that research should be done.

The survey respondents' comments in response to an open-ended question in our survey illustrate that research and small-scale testing are seen as ways to determine whether technologies can be safely and effectively deployed. In other words, respondents identified research and small-scale testing as ways to assess the potential for harm from climate engineering technologies and to allow for more informed decisions about their use.

The survey results also indicate that while approximately 65 percent of the public overall would support research on geoengineering, about half or more of the U.S. public across a range of demographic and political groups, including age, gender, race, ethnicity, education, and partisanship, would say that research should be done to determine whether geoengineering technologies are practical. In other words, support for research on geoengineering would not be limited to specific demographic groups.

To explore potential public support for government-sponsored research on geoengineering, our survey also asked respondents two separate questions about whether they would support or oppose the U.S. government's paying for research on CDR or SRM technologies. The responses to these questions suggest that if public information were similar to that in our survey, about half the public would support the U.S. government's paying for research on CDR technologies and about 45 percent would support its paying for research on SRM technologies.

As we remarked previously, public understanding of geoengineering is not well developed and our survey results do not necessarily predict future views. Furthermore, we did not ask respondents to consider the trade-offs between federal financing of geoengineering research and other possible spending priorities, including tax cuts or deficit reduction. We also did not ask respondents whether they supported private companies' or other entities' paying for research on climate engineering. Support for the government's paying for research on geoengineering technologies could have been less or more had we asked respondents to choose

alternative policy options or alternative funding sources. Research funded by a corporation or foreign government, for example, might yield different public support.

5.3 Views on climate engineering in the context of climate and energy policy

National surveys of U.S. public opinion have found broad public support for a variety of measures to increase energy efficiency and diversify the energy supply (Pew 2010, 3; Bittle et al. 2009, 11). To place the public's view of climate engineering in the broader context of public opinion on climate and energy policy, we asked survey questions about reducing CO_2 emissions by increasing reliance on noncarbon-based energy sources and other methods in addition to climate engineering. From the results, we estimate that about three-quarters of the public support (strongly support or somewhat support) developing more fuel-efficient cars, power plants, and other such technologies; encouraging businesses to reduce their CO_2 emissions; and relying more on wind and solar power (figure 5.1). About 65 percent of the public strongly or somewhat supports actions to encourage people to reduce CO_2 emissions—for example, by driving less or renovating their homes. At the same time, our results indicate that if the public were given the same type of information as in our survey, about half would strongly or somewhat support developing geoengineering technologies. About 45 percent strongly or somewhat support relying more on nuclear power.

As the Royal Society reported, concern has been raised that geoengineering proposals could reduce public support for mitigating the effects of

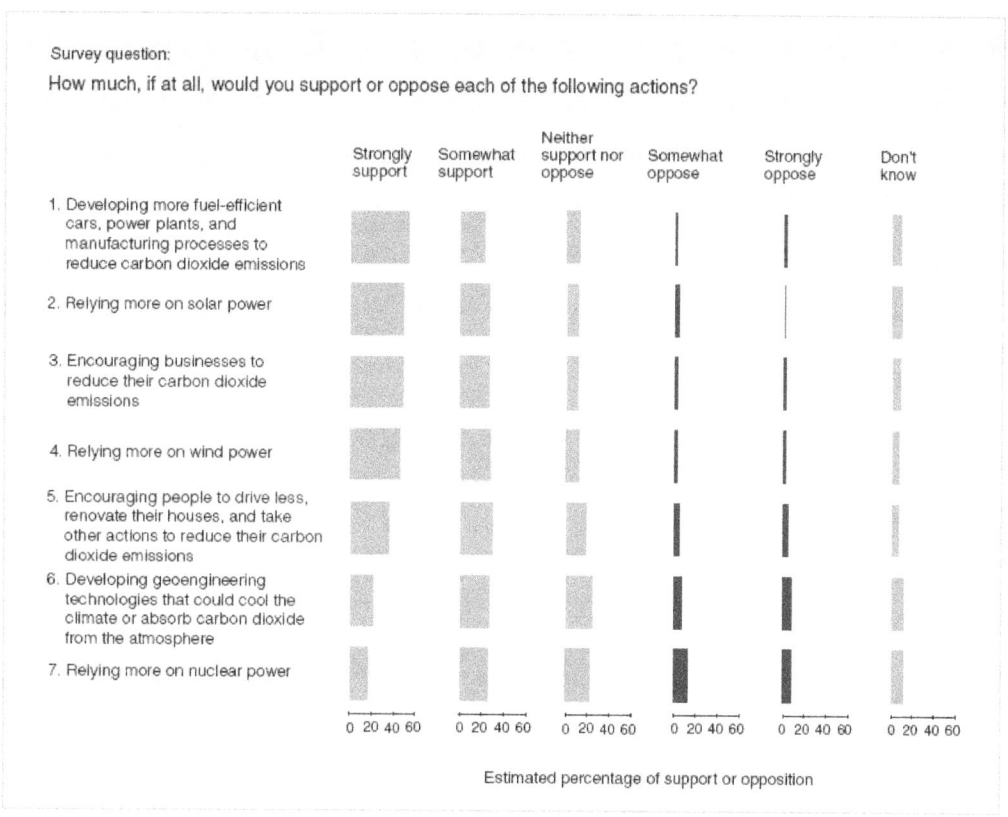

Survey question:

How much, if at all, would you support or oppose each of the following actions?

	Strongly support	Somewhat support	Neither support nor oppose	Somewhat oppose	Strongly oppose	Don't know

1. Developing more fuel-efficient cars, power plants, and manufacturing processes to reduce carbon dioxide emissions

2. Relying more on solar power

3. Encouraging businesses to reduce their carbon dioxide emissions

4. Relying more on wind power

5. Encouraging people to drive less, renovate their houses, and take other actions to reduce their carbon dioxide emissions

6. Developing geoengineering technologies that could cool the climate or absorb carbon dioxide from the atmosphere

7. Relying more on nuclear power

0 20 40 60   0 20 40 60   0 20 40 60   0 20 40 60   0 20 40 60   0 20 40 60

Estimated percentage of support or opposition

Figure 5.1 U.S. public support for actions on climate and energy, August 2010. Source: GAO.

Note: Estimates have 95 percent confidence intervals of within plus or minus 4 percentage points.

CO_2 emissions and could divert resources from adaptation (Royal Society 2009). This is referred to as the "moral hazard" problem. Given low public awareness of geoengineering, it is difficult to determine with any confidence whether the U.S. public would reduce support for mitigation as it learned more about geoengineering or how concerned the public would be about this moral hazard. Our survey results suggest that if the public were given the same type of information about geoengineering as our survey respondents, it might support a range of approaches to climate and energy policy, including climate engineering,

rather than viewing different approaches as trade-offs.

As with the results of qualitative research that found U.K. public support for combining geoengineering with mitigation efforts (Ipsos MORI 2010, 1–2), we found that at least some of the U.S. public views geoengineering as an additional method of addressing climate change rather than as an alternative to mitigation and adaptation. In open-ended comments, for example, some respondents expressed support for using other recognizable means to address climate change, such as reducing CO_2 emissions, and using geoengineering as a last resort.

5.4 Support for national and international cooperation on geoengineering

To obtain baseline information on U.S. public views on the extent to which different groups should be involved in deciding to use a geoengineering technology, our survey asked respondents how much involvement different public and private sector groups should have in making these decisions. From the results of our survey, we estimate that if the public were given the same type of information as in our survey, a total of about 75 percent would support a great deal, a lot, or a moderate amount of involvement by the scientific community in making decisions related to geoengineering (figure 5.2). At the same time, a total of about 70 percent would support a great deal, a lot, or a moderate amount of involvement by a coalition of national governments; about 65 percent would support this level of involvement by individual national governments, the general public, and private

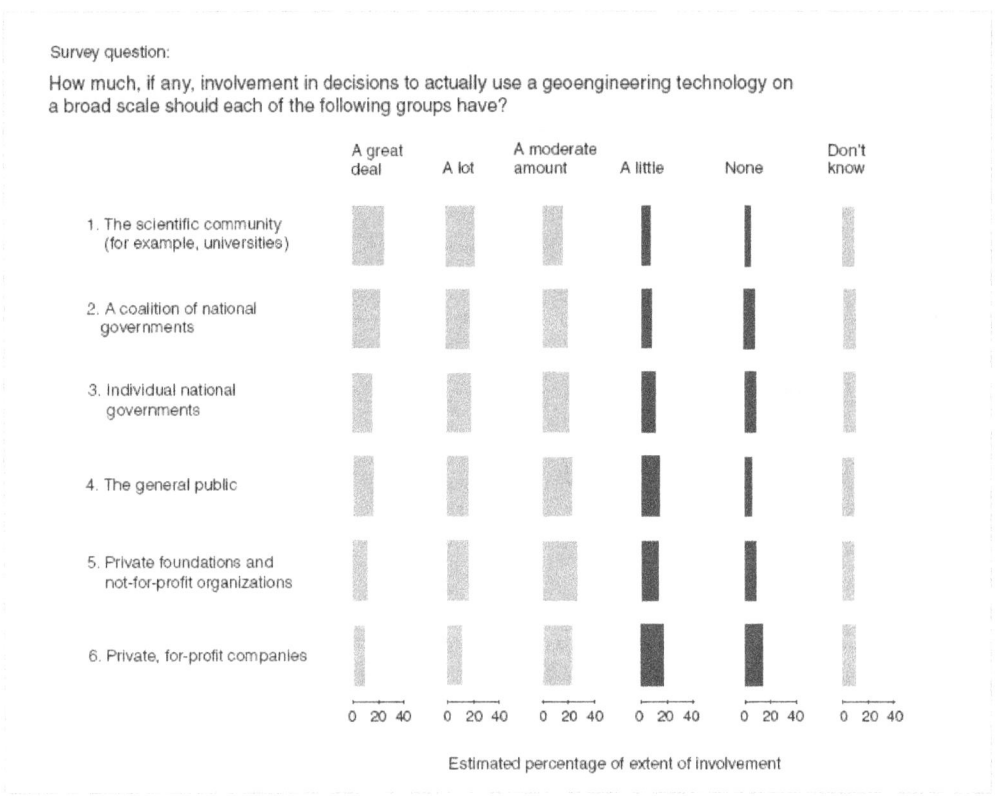

Figure 5.2 U.S. public views on who should decide geoengineering technology's use, August 2010. Source: GAO.

Note: Estimates have 95 percent confidence intervals of within plus or minus 4 percentage points.

foundations and not-for-profit organizations; and about half would support this level of involvement by private companies.

To provide additional insight into the U.S. public's initial views on actions related to geoengineering, we asked survey respondents whether they supported or opposed the U.S. government's coordinating more closely with other countries on geoengineering issues. We estimate that about 55 percent of the U.S. public would support the government's coordinating more closely with other countries on geoengineering issues, about 15 percent would oppose closer coordination, and about 30 percent would be unsure. Overall, the findings from our survey suggest that if the public were given similar information about geoengineering,

it would be open to the involvement of multiple national and international groups. In addition to expressing support for involvement by a range of groups in response to closed-ended questions, survey respondents noted the importance of involving the scientific community, governments, the public, and the private sector in making decisions about geoengineering in their answers to open-ended questions. In the words of one respondent,

"national governments, along with the scientific community, should determine under what circumstances it would be okay to actually use geoengineering technologies."

6 Conclusions

In this technology assessment, we have evaluated climate engineering technologies that could be part of a portfolio of climate policy options, along with mitigation and adaptation. We found that the technologies we reviewed are all in early stages of development. It is likely that significant improvements in climate engineering technology and related information will take decades of research because (1) today's technologies are not mature and (2) data collection and modeling capabilities related to climate engineering research are marked by important gaps. Experts have warned that a delay in starting significant climate engineering research could mean falling behind in our capacity to address a potentially damaging climate trend. We have previously reported that the United States does not have a coordinated strategy for climate engineering research.

We cannot ignore the possibility of new risks from either climate engineering research or its use or misuse. We found in our survey of U.S. adults that a majority would be open to climate engineering research but expressed concern about possible harm. Likewise, experts who advocate research emphasized that conducting significant climate engineering research and using the results could bring new risks, such as the possibility of international conflict arising from one nation's unilaterally deploying climate engineering technologies that adversely affect other nations. Additionally, future technological developments may bring new and currently unknown risks.

Experts we consulted suggested facilitating climate engineering researchers' interactions with the U.S. public, national decision-makers, and the international research community. They also said that international research could (1) help ensure that the nation is aware of and keeps pace with others' research and (2) give the United States an opportunity to lead by example by emphasizing transparency in and risk management for the research. Foresight efforts concerning emerging trends and technological developments could help the nation better anticipate future risks and opportunities.

7 Experts' review of a draft of this report

The fifteen experts listed in section 8.7 reviewed a draft of this report, at our request, and submitted comments to us. In this section, we summarize how we addressed the technical and other comments requiring a response. We also received a number of positive comments that do not require a response.

7.1 Our framing of the topic

Some comments pertained to the presentation of anthropogenic climate change or climate policy in the introduction to this report. These comments ranged from objecting to the presentation of alternative views of climate change to suggesting that we highlight scientific consensus on anthropogenic climate change. We retained information on the range of views as key introductory content but added a clarification acknowledging the endorsement of IPCC's view by numerous scientific bodies. We also added a statement linking the large consensus among authoritative scientific bodies to the sense of urgency that has contributed to discussions of engineering the climate. In response to other comments on the need to emphasize the potential role of climate engineering as a complement to mitigation and adaptation, we highlighted a GAO recommendation that the federal government develop a coordinated approach to geoengineering research in the context of a federal strategy to address climate change (GAO 2010a). We also incorporated suggestions to balance the risks introduced by engineering the climate against the risks of climate change without climate engineering.

7.2 Our assessment of technologies

Several comments surrounded the scope of the criteria we used to assess climate engineering technologies. Many of these comments concerned the appropriateness of TRLs to measure the readiness of soft climate engineering technologies for deployment, as opposed to devices or hard technologies. Given these comments, we discussed a key limitation of TRLs—that is, their sensitivity to certain criteria, such as the definition of a system concept or concrete plan. Developing an alternative way to measure the maturity of technologies was beyond our scope. In response to comments on the other key measurements, we revised the draft to emphasize the potential effectiveness and potential consequences of the technologies we assessed. To address concerns about the precision of cost estimates from the scientific literature, we focused on cost factors, or resources required to develop or deploy climate engineering technologies. Finally, we revised tables 3.1 and 3.2 to reflect these amplifications and clarifications.

7.3 Our assessment of knowledge and tools for understanding climate engineering

To incorporate comments on the status of knowledge and tools for understanding climate engineering, we reemphasized our focus on the value of research to help improve climate science, observational systems, or computing power. We replaced generalizations with examples of areas that scientists have targeted for improvements, and we strengthened our citations. We added climate chemistry models to our taxonomy of existing climate models. We updated some examples, such as NASA's CLARREO mission. We also accepted editorial comments clarifying certain ideas. For example, we characterized scientists' concerns about the reliability of observational networks in terms of the continuity of the observational record, and we revised the text to highlight the potential value of developing high-performance computing resources that could be dedicated to resolving uncertainties about regional climate variables. Although some comments noted that various observations could apply to other areas of climate science, we considered these comments to be beyond the scope of our report. Comments on decision-making under uncertainty were also beyond our scope.

7.4 Our foresight and survey methodologies

Comments on foresight and survey methodologies centered on the rationale for the content of the events described in the scenarios and the survey questions. In the foresight section of this report, the experts' views on the future are not based primarily on events described in the scenarios. Rather, the scenarios led a wide range of experts to share their views on the future of climate engineering research over the next 20 years (section 8.1.2), thus allowing a broad thematic discussion of these views, which sometimes differed sharply. The scenarios reflect the views of the experts who helped build them, but our overall foresight approach gave considerable latitude to the expression of views by all experts we consulted. Additionally, we included experts with many different kinds of expertise and varied views on climate engineering.

For these reasons, we are confident that our overall results would have been similar had the scenarios differed or been produced on the basis of an explicit underlying rationale.

Objectives for the survey included developing baseline information on public awareness of climate engineering technologies, views about research on them, and opinions on who should be involved in decisions related to climate engineering. Our focus groups and pretest interviews indicated that members of the public were unlikely to have either detailed knowledge or established opinions about climate engineering and that public views on climate engineering depended on the technology. Therefore, we developed and pretested (1) a basic definition of climate engineering with examples of different technologies, in both audio-visual and written formats so respondents could choose between the two, and (2) basic survey questions about each respondent's awareness of and views on research and groups that should be involved in decisions. Our pretest results led us to believe that the respondents understood the basic questions and that these were unbiased and provided the baseline information we needed to meet our objectives.

We did not incorporate other suggestions that were beyond the scope of our report. It was, for example, as much beyond our scope to develop a detailed strategy for deploying climate engineering as to compare a climate-engineered world with one lacking any deliberate climate intervention.

8 Appendices

8.1 Objectives, scope, and methodology

In this appendix, we describe the several targeted, coordinated methods we used to report on

- the current state of climate engineering technology,

- experts' views of the future of climate engineering research, and

- public perceptions of climate engineering.

In addition to the separate methods we used to address each objective, with the assistance of the National Academy of Sciences (NAS) we convened a meeting of scientists, engineers, and other experts that we called the Meeting on Climate Engineering. Because climate engineering is complex, NAS selected, with our assistance, a diverse and balanced group of experts on climate engineering, climate science, measurement sciences, foresight studies, emerging technologies, research strategies, and the international, public opinion, and public engagement dimensions of climate engineering. Experts participating in our Meeting on Climate Engineering are listed in section 8.6.

Before meeting in Washington, D.C., on October 6–7, 2010, the participants were provided with a written summary of our progress on this technology assessment. We explained to the participants that the summary was a working document showing what we had developed up to that point and that it did not fully describe our methodology.

The participants were organized into subgroups to focus on the major topics of our technology

assessment, including carbon dioxide removal (CDR) technologies, solar radiation management (SRM) technologies, the future of climate engineering, and public perceptions. The participants in each subgroup presented a 5-minute summary of their views on our preliminary findings, and then the entire group discussed the feedback. The meeting ended with general reactions to and advice and suggestions on our preliminary findings.

The participants' comments led us to review additional literature and unpublished studies that they suggested. Following the meeting, we also contacted the participants in person or by telephone or e-mail to clarify and expand what we had heard. We used what we learned from this meeting of experts to update, clarify, and correct where appropriate our information on the current state of climate engineering technology, expert views of the future of climate engineering research, and public perceptions of climate engineering. We incorporated in our draft report the lessons we learned from the meeting to give it greater accuracy and contextual sophistication.

8.1.1 Our method for assessing the state of climate engineering technology

To determine the current state of the science and technology of climate engineering, we reviewed a broad range of scientific and engineering literature. We started with the literature the Royal Society report referenced (Royal Society 2009, 63–68), and then we reviewed the literature we found in NAS, National Research Council (NRC), and U.S. government reports on climate change. We identified other literature from scientific and climate-related organizations such as the National Aeronautics and Space Administration (NASA) and National Oceanic and Atmospheric Administration (NOAA), and we reviewed proceedings from conferences such as the 2010 Asilomar International Conference on Climate Intervention Technologies (Asilomar Scientific Organizing Committee 2010). We revisited the report on climate engineering that we issued in September 2010 (GAO 2010a), which is complementary to this report. We reviewed relevant congressional testimony. We sought additional literature from the experts we spoke with.

We identified experts on climate engineering and proponents of specific climate change technologies from our review of the literature and conference proceedings. To ensure balance across the views and information we obtained, we interviewed a broad range of experts and officials working in climate science research and climate engineering whose track records had been proven through their peer-reviewed publications and presentations at conferences. We interviewed these experts to seek information that was not in their published work. We interviewed scientists, engineers, and knowledgeable officials with the Department of Energy's (DOE) National Energy Technology Laboratory (NETL), Lawrence Livermore National Laboratory, and Pacific Northwest National Laboratory (PNNL) and during site visits conducted at

- the National Center for Atmospheric Research,

- the National Oceanographic and Atmospheric Administration's Earth System Research Laboratory,

- the National Institute of Standards and Technology,

- Oak Ridge National Laboratory,

- American Electric Power's Mountaineer Power Plant in West Virginia,

- the Institute for Advanced Study at Princeton University,

- the Marine Biological Laboratory at Woods Hole,

- Scripps Institution of Oceanography, and

- Woods Hole Oceanographic Institution.

We interviewed selected attendees at the 2010 Asilomar International Conference on Climate Intervention Technologies.

We reviewed records of earlier interviews we had conducted on topics relevant to this technology assessment. We analyzed interviews with high-level private-sector officials from

- Alstom, which develops carbon capture technology and equipment;

- Dow, which conducts research on and development and manufacture of solvents or sorbents needed for CO_2 capture; and

- Schlumberger Carbon Services, which engages in geological mapping and the characterization of subterranean structures for storing CO_2.

We interviewed experts at academic institutions such as Columbia University, the Massachusetts Institute of Technology, and Stanford University.

Because climate science and climate engineering are interdisciplinary and extremely complex, with cross-cutting issues that may be beyond any one expert's realm, we synthesized information from an array of experts with diverse views on these subjects. We did not try to interview an equal number with alternative perspectives on all issues or technologies, because we were not evaluating the information we gathered by the number of experts who mentioned a topic or stated a particular view. Our objectives were to identify experts' (1) general understanding of the current state of climate science and engineering and (2) their major uncertainties and outstanding issues on these subjects. We did not attempt to determine the independence of individual experts, but we did try to obtain a balanced set of views. We wanted to obtain a broad perspective on the current state of climate science and engineering and objectively report this information. The experts we spoke with are listed in section 8.2.

We used the Royal Society's classification of climate engineering approaches to focus our review on CDR and SRM technologies (Royal Society 2009, l). We did not include climate engineering approaches that address other, non-CO_2 greenhouse gas emissions such as nitrous oxide. After consulting with experts, we limited our assessment of climate engineering technologies to those the Royal Society addressed in its 2009 report. Of those technologies, we did not assess ocean reflectivity or ocean upwelling or downwelling because we found limited information on them in the peer-reviewed literature. We also did not assess research on the possible causes of climate change.

We assessed and described the current status of climate engineering technologies along four key dimensions: (1) maturity, (2) potential effectiveness, (3) cost factors, and (4) potential consequences. We assessed the maturity of climate engineering technologies by their technology readiness levels (TRL) (table 8.1). TRLs are a standard tool for assessing the readiness of an emerging technology for production or incorporation into an existing technology or system. The Department of Defense and NASA use TRLs, as does the European Space Agency.

We used the AFRL (Air Force Research Laboratory) Technology Readiness Level Calculator Version 2.2 (Nolte 2004) to determine technology readiness levels for the climate engineering technologies we reviewed. Table 8.1 summarizes key features of TRL ratings. The first column presents definitions of TRL levels used as "Top Level Views" in the TRL calculator. The calculator operates conditionally: to achieve a rating at any level, a technology must satisfy the requirements for all lower levels as well. For example, to achieve a rating of TRL 2, a technology must also satisfy the requirements for a rating of TRL 1. To achieve a rating of TRL 3, a technology must also satisfy the requirements for a rating of TRL 2, and thus must also satisfy the requirements for a rating of TRL 1.

We developed criteria to rate climate engineering technologies using the TRL calculator. For the top level view of TRL 1, requiring that basic principles be observed and reported, we asked whether the technology had been described as a climate engineering technology in peer-reviewed scientific literature. All the climate engineering technologies we reviewed met this condition.

For the top level view of TRL 2, requiring the formulation of a technology concept or application, we asked whether a system concept identifying key elements of the technology or a concrete plan existed for implementation on a global scale. Some technologies failed to meet this condition for climate engineering even though they would be fully mature in other applications. For example, increasing the reflectivity of settled

Level	Description	Example
1 Basic principles have been observed and reported	The lowest level of technology readiness. Scientific research begins translation into applied research and development	Paper studies of the technology's basic properties
2 Technology concept or application has been formulated	Invention begins. Once basic principles are observed, practical applications can be invented. The application is speculative, and no proof or detailed analysis supports the assumption	Limited to paper studies
3 Analytical and experimental critical function or characteristic proof of concept has been defined	Active research and development (R&D) begins. Includes analytical and laboratory studies to physically validate analytical predictions of separate elements of the technology	Components that are not yet integrated or representative
4 Component or breadboard validation has been made in laboratory environment	Basic technological components are integrated to establish that the pieces will work together. This is relatively "low fidelity" compared to the eventual system	Ad hoc hardware integrated in a laboratory
5 Component or breadboard validation has been made in relevant environment	Fidelity of breadboard technology increases significantly. The basic technological components are integrated with reasonably realistic supporting elements so the technology can be tested in a simulated environment	"High fidelity" laboratory integration of components
6 System and subsystem model or prototype has been demonstrated in a relevant environment	Representative model or prototype system is well beyond level 5 testing in a relevant environment. Represents a major step up in the technology's demonstrated readiness	Prototype tested in a high-fidelity laboratory or simulated operational environment
7 System prototype has been demonstrated in an operational environment	A prototype is operational or nearly operational. Represents a major step up from level 6, requiring the demonstration of an actual system prototype in an operational environment, such as in an aircraft, vehicle, or space	Prototype tested in a test bed aircraft

Table 8.1 Nine technology readiness levels, continues on next page

Level	Description	Example
8 Actual system is complete and has been qualified in testing and demonstration	Technology has been proven to work in its final form and under expected conditions. In almost all cases, this level represents the end of true system development	Developmental test and evaluation of the system to determine if it meets design specifications
9 Actual system has been proven in successful mission operations	The technology is applied in its final form and under mission conditions, such as those encountered in operational test and evaluation. In almost all cases, this is the end of the last "bug fixing" aspects of true system development	The system is used in operational mission conditions

Table 8.1 Nine technology readiness levels described. Source: GAO based on Nolte (2004).

Note: A breadboard is a representation of a system that can be used to determine concept feasibility and develop technical data. It is typically configured for laboratory use only. It may resemble the system in function only.

areas by painting rooftops white would be mature on a small scale but lacked a system concept and a concrete plan for implementation on a global scale. Since this technology failed to meet the condition for TRL 2, it was rated at TRL 1. The sensitivity of the TRL ratings to the definition of a system concept or a concrete plan for climate engineering is a key limitation of using TRLs to evaluate technologies that are otherwise mature.

For the top level view of TRL 3, requiring analytical and experimental demonstration of proof of concept, we looked for significant experimental data on elements of the technology. For example, a technology designed to reduce solar radiation by placing scatterers at L1 fulfilled the basic requirements for TRL 2 but not TRL 3 because the supporting literature was theoretical and did not provide experimental data. Finally, for the top level view of TRL 4, requiring technological demonstration, we looked for evidence of system demonstration with a

breadboard unit (a representation of the system, in function only, used to determine feasibility and to develop data, configured for laboratory use). Because none of the technologies that we reviewed had system data with breadboard units, none could be rated at TRL 4 or higher.

We had earlier recommended that a technology should be at level 7—that is, a prototype has been demonstrated in an operational environment—before being moved to engineering and manufacturing development. We had recommended further that a technology be at level 6 before starting program definition and risk reduction (GAO 1999). We characterize technologies whose TRL scores are below 6 as "immature."

Two factors that affect global temperature are (1) the level of CO_2 and other greenhouse gases in the atmosphere and (2) the amount of solar radiation that Earth and its atmosphere absorb.

Because CDR and SRM affect temperature in different ways, their effects are measured differently. CDR removes CO_2 from the atmosphere while SRM reduces the amount of solar radiation that Earth and its atmosphere absorb—by reflecting the radiation into space before it reaches Earth's atmosphere, when it reaches Earth's atmosphere, or when it reaches Earth's surface.

To describe the effectiveness of proposed CDR technologies, we examined the Royal Society's qualitative ratings of various technologies' effectiveness (high, medium, and low). We also examined two quantitative measures reported in the literature: the estimated (1) maximum reduction of the atmospheric concentration of CO_2 (ppm) from its projected level of 500 ppm in 2100 and (2) annual ability to remove CO_2 from Earth's atmosphere (gigatons of CO_2 per year) when compared to annual anthropogenic emissions of 33 gigatons of CO_2.[88] We assessed the qualitative ratings primarily by making

- a check for reasonableness. For example, for bioenergy with CO_2 capture and sequestration (BECS) the Royal Society reported an anticipated maximum CO_2 reduction ability of between 50 ppm and 150 ppm and rated BECS as low to medium in effectiveness. We confirmed the reasonableness of rating a reduction of 150 ppm as having medium effectiveness by noting that this level of reduction would put the concentration of CO_2 in the year 2100 at 350 ppm—which is below the current 390 ppm but does not approach the preindustrial 280 ppm.

- comparisons to other scientific sources. We reviewed scientific literature for other assessments indicating the overall feasibility of implementing individual CDR technologies on a global scale to achieve a net reduction of atmospheric CO_2 concentration. For two technologies—direct air capture of CO_2 with geologic sequestration and enhanced weathering—sources in the peer-reviewed literature provided views or information that differed substantially from the Royal Society's ratings.[89]

Overall, for three of the six CDR technologies, our assessments confirmed the specific Royal Society qualitative effectiveness ratings. We included these three Royal Society ratings in the "potential effectiveness" column in table 3.1. For one other technology (land use management), which the Royal Society rated as low, other scientific literature suggested a low to medium rating, which is reflected in table 3.1. For the remaining two CDR technologies (direct air capture of CO_2 with sequestration, and enhanced weathering), we did not report an overall qualitative rating for potential effectiveness; that is, we indicated "not rated" because sources in the scientific literature provided information that differed considerably from the Royal Society's ratings. However, where possible, we provided other relevant information.

To describe the potential effectiveness of SRM technologies, we used the generally accepted benchmark of the climate change community (such as in the work of the Intergovernmental Panel on Climate Change (IPCC)) called

[88] The preindustrial CO_2 concentration is reported to have been 280 ppm. In 2010, the atmospheric concentration of CO_2 was estimated in the literature as 390 ppm. In the year 2100, the concentration projected for a mitigation scenario is 500 ppm.

[89] The high effectiveness rating the Royal Society gave for these two technologies could not be confirmed and validated by reports in the literature. We did not assign an overall qualitative rating to these technologies because of conflicting indications in the literature about their effectiveness.

equilibrium climate sensitivity.[90] Climate modeling studies use equilibrium climate sensitivity as a benchmark to indicate the effect of greenhouse gases on the climate. Equilibrium climate sensitivity is defined as the change in global mean surface temperature following warming caused by a doubling of preindustrial CO_2 levels (Solomon et al. 2007). The doubling of preindustrial CO_2 levels is also used in modeling studies as a standard condition for evaluating climate effects other than an increase in global average temperature. Following this approach, the climate engineering community evaluates the effects of SRM technologies against double preindustrial CO_2 levels. We described the potential effectiveness of SRM technologies when fully implemented on a global scale, based on the extent to which they are estimated to reduce global average surface temperature compared to the benchmark. We categorized the potential effectiveness of each climate engineering technology as a percentage, where 100 percent is anticipated to lower global mean temperature from the benchmark to the preindustrial value and is termed "fully effective."

We did not assess the effectiveness of either deploying multiple climate engineering technologies simultaneously or combining them with reductions in carbon emissions and advances in energy technology. We did not assess the effectiveness of deploying a technology in any specific place. Because we focused on global mean surface temperature, we did not assess specific geographic temperatures or climate changes.

We did not independently determine the costs of implementing the technologies. Instead, we report cost factors and estimates from the literature we reviewed; these are based on ideas of what the technologies might be, not on detailed design and schedule data. For CDR, the cost factors represent resources used to remove CO_2 from the atmosphere and store it; when quantified, these are presented on a per ton basis. For SRM, the cost factors represent resources required to counteract global warming from doubling the preindustrial atmospheric concentration of CO_2, or, for technologies that are not anticipated to be fully effective, the resources required to counteract warming to the maximum extent possible. We were not able to determine the reliability of estimated costs in the literature because of insufficient data or inadequate descriptions of how costs were determined.

We assessed the potential consequences of each technology by summarizing risks or consequences identified in the literature, modeling studies, and our interviews with experts. We also reviewed congressional hearings for the testimony of experts who presented risks of implementing specific technologies. We reviewed the ability of existing climate models to represent climate processes expected to result from climate engineering technologies, including altered wind currents, rain patterns, and ocean temperatures. We considered the ability to reverse a technology's deployment as a type of consequence.

To report on the status of knowledge and tools for understanding climate engineering, we reviewed relevant literature and interviewed scientists and other experts about climate science, observational networks, and computing resources. Our literature review included GAO publications as well as reports from the Department of Energy, Intergovernmental Panel on Climate Change, National Aeronautics and Space Administration,

[90] The word "equilibrium" indicates a steady state response to specify climatic conditions, such as the concentration of CO_2 and variables related to climate engineering.

National Institute of Standards and Technology, National Oceanic and Atmospheric Administration, National Research Council, United States Climate Change Science Program, and World Climate Change Programme, in addition to peer-reviewed scientific literature. Because we found few studies focusing on climate engineering modeling or research, we included in our review some studies of climate models and science that are relevant to climate engineering. Our research on observational systems focused on the coverage, continuity, and accuracy of networks collecting measurements related to substances or processes that are important to climate engineering. Similarly, our examination of computing resources focused on current limitations or potential improvements that could affect climate engineering research, such as the spatial resolution of computations in current models. We did not independently evaluate whether scientific knowledge or tools are sufficiently well understood or developed for making decisions about the possible development or use of climate engineering technologies. We also did not assess whether climate change is occurring or what is causing any climate change if it is occurring or whether current scientific knowledge supports the occurrence of climate change or its causes.

8.1.2 Our method for eliciting experts' views of the future of climate engineering research

To assess how climate engineering research might develop in the future, we used the following three sources: (1) a foresight exercise in which experts developed alternative scenarios, (2) the comments of a broad array of experts stimulated by the scenarios, and (3) additional views of other experts in response to our preliminary synthesis developed from the scenarios and

earlier comments. Sections 8.4 to 8.6 list the experts we consulted in developing each of these sources.[91] We present our summary of the three sources in the body of our report to suggest some possibilities for climate engineering research over the next 20 years.

All experts we selected to participate in the foresight exercise and to comment in response to the scenarios met at least one of the following criteria: they (1) held a position in a university or other well-known organization relevant to climate engineering, climate change, or related topics; (2) had participated in academic or professional panels addressing climate engineering, climate change, or related topics; or (3) had authored peer-reviewed publications on climate engineering, climate change, or related topics.

8.1.2.1 Scenario-building process

A meeting to build scenarios held on July 27, 2010, at GAO headquarters was facilitated by a professional from the Institute for Alternative Futures. The overall goal of the exercise was to develop four scenarios to illustrate alternative possible futures for climate engineering research, including the amounts and kinds of research that might be conducted on CDR and SRM and whether significant progress was expected.

The scenario-builders (listed in section 8.4) were selected to constitute, as a group,

- expertise on specific technologies for engineering the climate, including CDR and SRM, and experience in the research or development of relevant technologies;

---

[91] Additionally, in preparing for these activities we interviewed other experts who provided background information or recommended some of the experts listed in sections 8.4 to 8.6.

- knowledge on climate engineering as well as the development of future-oriented scenarios, including foresight about emerging technologies and national and international approaches to them; and

- collective backgrounds in private industry, government (including the military), and other organizations such as those in academia.

We selected six external scenario-builders. Each of the six was a leading expert in one or more key fields or had been recommended to us by other experts. The group's knowledge and expertise represented a balance across the items bulleted above and spanned energy policy, climate change, oceanography, atmospheric science, and biotechnology, as well as research on CDR and SRM and other areas, such as foresight and public engagement. Timothy Persons, GAO's Chief Scientist, served as the host and ex-officio member of the group to help guide the discussion.

To build the four scenarios, we began by reviewing scientific and engineering literature and interviewing scientists and engineers to help us identify what were likely to be the key factors in the future scope and direction of climate engineering research in the United States. We used this information to construct a questionnaire that we sent by e-mail before the meeting to the six external experts and GAO's Chief Scientist.

Before the scenario-builders met on July 27, 2010, they responded individually to our e-mail questionnaire. The questionnaire asked for their opinions on the goals of climate engineering research, the importance of making substantial progress toward those goals by 2030, the promise of different approaches toward reaching the goals, the research that might appropriately be supported by private or government funds, any leadership the federal government should take on climate engineering research, the need for international cooperation, the likelihood of future climate changes, and the moral hazard if climate engineering research looked as if it were headed on an efficient and effective course. We also asked for separate answers to these questions as they related expressly to CDR and SRM technologies. The questionnaire also listed various factors that might affect climate engineering research and asked for the scenario-builders' opinions on these and other relevant factors. The answers we received suggested the importance of factors subsequently selected for the meeting's discussion. For example, five of the six scenario-builders responded that government incentives to industry would make the prospect of achieving some or all CDR research goals by 2030 "highly promising." We provided them with a summary of their answers to the questionnaire at the outset of the scenario-building meeting.

During the meeting, the scenario-builders identified and discussed many kinds of factors important for future U.S. climate engineering research in a global context. They selected two policy-related factors as potentially most significant. One was whether a federal research program on CDR and SRM would be established and, if so, at what level (the scenario-builders did not focus on low-risk SRM methods such as whitening roofs and roads). The scenario-builders discussed a broad definition of a research program that might include related activities, such as engaging the public or encouraging industry to implement technology-related results (including improving opportunities for dissemination). The other policy factor was whether carbon constraints would be established and, if so, at what level in the United States and internationally.

The scenario-builders discussed how carbon constraints can take the form of either emissions pricing or regulations designed to reduce carbon emissions. After selecting the two factors—a federal research program and carbon constraints—the scenario-builders specified three levels for each one, defining nine combinations, each of which might serve as the basis for a scenario (figure 8.1). From the nine possible combinations, the scenario-builders selected for further consideration the four combinations labeled on figure 8.1. The four resulting scenarios define a range of futures within the bounds set by variation across the two selected factors.

Each scenario was developed separately for a specific combination of factors. However, a logical inference is that more pathways are possible within the range defined by the two factors because of the possibility of transitions from one scenario to another. Scenario II, for

example, could overlap with Scenario IV. The purpose of the scenarios was to stimulate thinking about the future, not to limit anticipation to any one cell.

We asked the scenario-builders to identify low-probability high-impact events such as "black swans" and "black pearls." We defined black swan as an extremely unlikely event able to produce catastrophic or otherwise large effects. We defined black pearl as a black swan with positive effects. We generated this list to help identify wild cards or conditions that could drastically change the future as related to the climate and climate engineering research.

Toward the end of the scenario-building meeting, we asked the six external scenario-builders to look ahead to 2030 and beyond and to consider possible outcomes linked to research on CDR and SRM. We asked them to assess, subjectively and qualitatively, three potential future

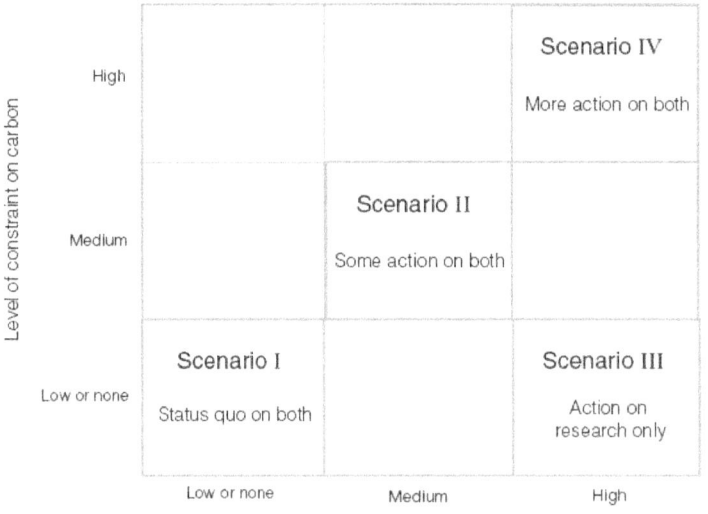

Figure 8.1 Four scenarios defining alternative possible futures. Source: GAO.

situations that might occur in or after 2030: (1) an emergency in which decision-makers might consider using SRM, (2) continued global warming, and (3) a future with no further warming.

8.1.2.2 Experts' comments stimulated by scenarios

We used the scenarios to elicit additional views about the future from 28 experts (listed in section 8.5) who represented a wider range of backgrounds and perspectives. To help ensure balance in the wider group of experts who would review and respond to the scenarios, we specifically selected some experts with competing views and different backgrounds. These experts were thus characterized by

- varied backgrounds (including, for example, economics, ethics, the humanities, and international relations);

- a range of organizational affiliations (including universities, the public sector, the private sector, and advocacy groups or other organizations associated with a viewpoint); and

- differing perspectives (including some known to favor or oppose the development of climate engineering technologies or to have expressed uncertainty about climate change trends).

In August 2010, we e-mailed the four scenarios to the selected 28 experts along with a brief questionnaire on their reactions to the scenarios. We invited them to provide alternative mini scenarios or other statements of their views about the future. We asked them to identify black swans and black pearls. We also asked them for any message about the future of climate engineering research and its consequent risks that they believed would be important

for policymakers to consider. Not all expressed views on all issues. We followed up with e-mail questions for clarification, as needed. In a few instances, we followed up with telephone conversations or met in person with experts who were available in the Washington, D.C., area. We synthesized the varied responses we received from the experts.

8.1.2.3 Experts' views of our initial synthesis and preliminary findings

As we described above, we convened with NAS's help a meeting of scientists, engineers, and other experts. For this meeting, we presented information about the scenarios and asked the experts to discuss our preliminary findings about views expressed regarding the future and to share their own views about the future. Some experts did not express views on the scenarios or all topics discussed.

8.1.2.4 Our analysis: A qualitative foresight synthesis

We call our summary of the combined results of the exercises we have described a qualitative foresight synthesis. The summary is primarily based not on how many experts made specific comments or any number of votes taken of the experts but, rather, on a qualitative approach in which we identified recurring, prominent themes and used professional judgment. The summary is a synthesis of views from a diverse range of experts and from three interconnected foresight exercises. It is the result of an iterative process whereby one set of experts developed scenarios, another set commented on those scenarios, and a third set reviewed our initial synthesis of the first two exercises. In areas where either a clear majority of the experts we consulted agreed or only a small number took a specific position, we say that a "majority" expressed the position or that a small number stated a concern. However, for transparency, footnotes provide information on specific counts of experts we consulted who voiced key opinions.

Although the experts we consulted do not necessarily represent the views of all those with similar expertise in the area of climate engineering, because of the three-stage process and the breadth of experts we consulted, we believe that the resulting overall set of views about the future that we present in section 4 of this report ("Experts' Views of the Future of Climate Engineering Research") would be similar even if we had used a different set of scenarios or if we had consulted with a different but still diverse set of experts.

8.1.3 Our method for assessing potential responses to climate engineering

To gather information about public awareness of and views on geoengineering technologies, we reviewed selected survey research on public opinion on climate change, conducted focus groups, and contracted with Knowledge Networks Inc. to use its online research panel to field a survey we developed. The survey was fielded from July 19 to August 5, 2010. Of a total sample of 1,623 U.S. residents 18 years old and older, 1,006 completed the survey.

From our review of the research on climate engineering and survey research on climate change, we did not expect the focus group participants or survey respondents to know very much about climate engineering technologies. Therefore, before asking questions about geoengineering, we gave the focus group participants and the survey respondents a basic definition of geoengineering, described

the differences between CDR and SRM, and provided examples of both. The level of information we gave the focus group participants and survey respondents was comparable to what average adults exposed to news media descriptions of these technologies might be expected to receive.

To help us develop the protocol for the focus group with members of the public and to increase our understanding of public perceptions of climate change and climate engineering, we conducted four focus groups with GAO employees. We used what we learned from these focus groups to make changes to the protocol for the public focus group.[92] We selected the 11 members of the public focus group for their diversity in age, gender, race, ethnicity, and education. Some participants spoke both English and Spanish; they translated for one participant who was fluent only in Spanish. A GAO analyst fluent in English and Spanish observed the focus group.

We first asked the focus group participants to discuss their beliefs about climate change, including whether they believed the climate is changing and, if so, what the cause is. We then asked them if they thought there was anything they personally could do to affect climate change and what, if anything, the public, industry, government, or scientists and engineers should do with respect to climate change. Participants identified personal actions such as driving less, using alternative fuels, and writing letters to influence elected representatives. With respect to government, industry, and scientists

[92] We also conducted two focus groups with science and engineering graduate students participating in Arizona State University's Consortium for Science, Policy & Outcomes (CSPO), one before and one after the public focus group. We did not make any changes to the focus group protocol as a result of the CSPO focus group conducted before the public focus group.

and engineers, participants thought greater enforcement of existing laws, the provision of government incentives to address climate change, and increased public education about climate change were ways to address climate change. When asked whether they were aware of any scientific or engineering solutions to climate change, focus group participants did not identify any specific solutions. One participant stated that scientists and engineers might develop solutions to climate change but that money is not being directed to this.

After asking focus group participants if they were aware of any scientific or technological solutions to climate change, we explained what geoengineering is and gave them information about three different technologies, including CDR and SRM technologies. We asked participants to discuss their reactions to each technology and whether they supported or opposed it. In addition, we asked them to discuss how the federal government, industry, and individuals should fund and make decisions about geoengineering.

We chose to use "geoengineering" in the information we gave the focus group and survey participants, given that we and others, such as the Royal Society, had used this term earlier. In our focus groups, we found that participants raised concerns about the potential for harm from geoengineering technologies and reacted differently to different technologies. For example, one participant, asked to react to information about stratospheric sulfates, expressed the view that dinosaurs had become extinct by the Sun's having been blocked. Another, reacting to the concept of direct air capture, expressed concern about the long-term storage of CO_2.

To assess whether these differences in reaction to different technologies exist also in the larger

population, we administered a split-ballot survey. Using experts' assessments of safety described in the Royal Society report on geoengineering, we gave half the respondents information about technologies (one CDR and one SRM) that experts identified as relatively safe, and we gave the other half information about technologies (one CDR and one SRM) that experts identified as relatively less safe. We included a question in the survey to assess whether this difference in information about experts' assessment of safety affected participants' perceptions of potential harm from CDR and SRM technologies. We also examined whether views about geoengineering research, development, and decision-making were affected by learning about more or less safe technologies. Our survey results indicated that respondents differed in their level of concern about harm from geoengineering, depending on whether they received information about more or less safe technologies, but they did not differ greatly in their responses to other questions about geoengineering research and decision-making. Consequently, we report the results from all other survey questions combined.

The respondents could choose one or more of three ways to receive information about different types of geoengineering technologies: they could (1) view a video and listen to a narration, (2) listen to the narration, or (3) read printed information. All survey questions were identical in the two survey ballot groups.

Every survey introduces sampling and nonsampling errors, including errors of processing, measurement, coverage, and nonresponse. We took steps to reduce such errors. To reduce processing error, we verified all computer programming and analyses independently. To reduce measurement error, we conducted 11 pretests with persons of varied education, income, English proficiency, age, gender, and race.[93] The pretests included face-to-face interviews using the draft written survey as well as telephone interviews with those completing the web version of the survey. From the pretest results, we made a number of changes to reduce the likelihood of measurement error from respondents' misunderstanding or misinterpreting the survey questions. We also asked all pretest respondents whether any specific questions or the survey overall was biased in any way, and we made changes to address the concerns they raised. Knowledge Networks' online research panel was designed to minimize errors of coverage of the target population of U.S. adults. The sample frame was based on probability sampling that covered both people who had home access to the Internet and those who did not. Knowledge Networks also used a dual sampling frame that included both households that had telephones (including only cell phones) and households that did not, as well as households with listed and those with unlisted telephone numbers. Knowledge Networks recruited panel members randomly. Households were provided with access to the Internet and the necessary hardware if they needed it. For a specific survey like ours, Knowledge Networks selects panelists randomly, and no one not selected may respond.

To calculate the survey's response rate, we used RR4, a method described by the American Association for Public Opinion Research. The RR4 method is based on multiplying the recruitment rate (18.3 percent), the profile rate (58.4 percent), and the completion rate (62.0 percent) to yield an overall response rate of 7 percent. To reduce the potential for nonresponse error, we weighted the survey data using Knowledge Networks' study-specific post-stratification weight. From our assessment of Knowledge Networks' probability sampling methods and weighting methodologies and the results of our nonresponse bias analysis, we determined that the sample selected for our study was statistically representative of the U.S. adult population.

Sampling error is a measure of the likely variation introduced in a survey's results by using a probability procedure based on random selections. In terms of the margin of error at the 95 percent confidence level, the sampling error for survey estimates from the total sample is plus or minus 4 percentage points, unless otherwise noted. In terms of the margin of error at the 95 percent confidence level, the sampling error for estimates based on subgroups of the sample is plus or minus 9 percentage points, unless otherwise noted. Because the overall response rate was low and because sources of nonsampling error such as differences in survey results from panel attrition and panel conditioning might be present, nonsampling error may also have contributed to the total survey error of the results. To avoid false precision, therefore, we rounded the survey results we report in the text to the nearest 5 percentage points.

The public perceptions elicited by this survey are based on limited information about geoengineering and do not necessarily predict U.S. public views. We found that about 65 percent of the respondents had not heard about geoengineering before reading the survey; therefore, responses to the survey are likely to reflect reactions to information about geoengineering that we provided in the survey.

---

[93] Before we pretested our survey, students in the Science, Technology, and Public Policy Program at the Gerald R. Ford School of Public Policy, University of Michigan, provided input on issues related to governance and surveying public opinion in the area of climate engineering.

If the respondents had been provided with different information about geoengineering, the survey responses could also differ. Also, climatic or other events might change public views of geoengineering. When we asked respondents about their support for geoengineering research or for government funding of geoengineering research, we did not present them with competing programs to choose from (programs for cancer treatment, for example) or with alternatives, such as using government funding for national defense or cutting taxes. These kinds of choices might have produced different results.

The initial version of the survey included a question designed to help assess whether the respondents thought that exploring geoengineering solutions could distract from other potential solutions to climate change, such as reducing CO_2 emissions by driving less or developing more fuel-efficient technologies. Because the question included more than one policy option on which respondents could hold different views and focused on what respondents would expect to happen in the future but could not yield direct information about how members of the public might actually behave, we revised the survey to include a separate series of questions to assess where initial support for geoengineering might fall relative to other policy options.

8.1.4 External review

We invited all participants in the Meeting on Climate Engineering to review our draft report. We sent the draft report to the 16 participants who agreed to review and help revise the report. While we asked the 16 reviewers to focus on the sections most relevant to their expertise, we also told them that we welcomed any comments on the entire draft. One of the 16 did not participate in the review because of schedule conflicts. Fifteen reviewers (see section 8.7)

provided technical or other comments that we incorporated as appropriate. These 15 reviewers were meeting participants who collectively represented expertise relevant to each of the three major areas of our report, including the current state of climate engineering technology, expert views of the future of climate engineering research, and public perceptions of climate engineering. The external review was conducted in February 2011.

We conducted our work for this technology assessment from January 2010 through July 2011 in accordance with GAO's quality standards as they pertain to technology assessments. Those standards require that we plan and perform the technology assessment to obtain sufficient and appropriate evidence to provide a reasonable basis for our findings and conclusions based on our technology assessment objectives and that we discuss limitations of our work. We believe that the evidence we obtained provides a reasonable basis for our findings and conclusions, based on our technology assessment objectives.

8.2 Experts we consulted on climate engineering technologies

Barrett, Scott, Lenfest-Earth Institute Professor of Natural Resource Economics, School of International and Public Affairs and Earth Institute, Columbia University, New York.

Benford, Gregory, Professor of Physics, Department of Physics and Astronomy, University of California, Irvine.

Caldeira, Ken, Physicist and Environmental Scientist, Energy and Environmental Sciences Directorate, Lawrence Livermore National Laboratory, Livermore, California.

Crutzen, Paul J., Emeritus, Max Planck Institute for Chemistry, Mainz, Germany; Institute Scholar, International Institute for Applied Systems Analysis, Laxenburg, Austria; Emeritus Professor, Scripps Institution of Oceanography, University of California at San Diego, La Jolla.

Doney, Scott C., Senior Scientist, Department of Marine Chemistry and Geochemistry, Woods Hole Oceanographic Institution, Woods Hole, Massachusetts.

Ducklow, Hugh W., Director and Senior Scientist, The Ecosystems Center, Marine Biological Laboratory, Woods Hole, Massachusetts; Professor, Department of Ecology and Evolutionary Biology, Brown University, Providence, Rhode Island.

Dyson, Freeman, Professor Emeritus, School of Natural Sciences, Institute for Advanced Study, Princeton, New Jersey.

Fahey, David W., Research Physicist, Atmospheric and Chemical Processes, Chemical Sciences Division, Earth System Research Laboratory, National Oceanic and Atmospheric Administration, Boulder, Colorado.

Garten, Jr., Charles T., Senior Research Staff Member, Nutrient Biogeochemistry Group, Environmental Sciences Division, Oak Ridge National Laboratory, Oak Ridge, Tennessee.

Gibbons, John H. (Jack), President, Resource Strategies, The Plains, Virginia; Consultant, Lawrence Livermore National Research Laboratory, Livermore, California; Division Advisor, Division on Engineering and Physical Sciences, The National Academies, Washington, D.C.

Hack, James J., Director, National Center for Computational Sciences, Oak Ridge National Laboratory, Oak Ridge, Tennessee.

Keeling, Ralph, Professor, Scripps Institution of Oceanography, University of California at San Diego, La Jolla.

Keith, David, Director, ISEEE Energy and Environmental Systems Group; Professor and Canada Research Chair of Energy and the Environment; Professor, Department of Chemical and Petroleum Engineering, University of Calgary, Calgary, Alberta, Canada; Adjunct Professor, Department of Engineering and Public Policy, Carnegie Mellon University, Pittsburgh, Pennsylvania.

Lackner, Klaus, Department Chair, Ewing and J. Lamar Worzel Professor of Geophysics, Earth and Environmental Engineering and Director, Lenfest Center for Sustainable Energy, The Earth Institute, Columbia University, New York.

Latham, John, Emeritus Professor of Physics, University of Manchester, United Kingdom; Visiting Professor, University of Leeds, United Kingdom; and Senior Research Associate, National Center for Atmospheric Research, Boulder, Colorado.

Lindzen, Richard S., Alfred P. Sloan Professor of Meteorology, Department of Earth, Atmospheric, and Planetary Sciences, Massachusetts Institute of Technology, Cambridge, Massachusetts; Distinguished Visiting Scientist, Jet Propulsion Laboratory, California Institute of Technology, Pasadena, California.

Long, Jane C. S., Associate Director, Energy and Environment Directorate, Lawrence Livermore National Laboratory, Livermore, California.

MacCracken, Michael, Chief Scientist for Climate Change Programs, Climate Institute, Washington, D.C.

MacDonald, Alexander E. "Sandy," Deputy Assistant Administrator for Laboratories and Cooperative Institutes, Office of Oceanic and Atmospheric Research; Director, Earth System Research Laboratory, National Oceanic and Atmospheric Administration, Boulder, Colorado.

Marland, Gregg, Distinguished R&D Staff, Environmental Sciences Division, Oak Ridge National Laboratory, Oak Ridge, Tennessee; Guest Professor, Ecotechnology Program, Mid Sweden University, Östersund, Sweden.

Melillo, Jerry M., Distinguished Scientist, The Ecosystems Center, Marine Biological Laboratory, Woods Hole, Massachusetts; Professor (MBL) of Ecology and Evolutionary Biology, Department of Ecology and Evolutionary Biology, Division of Biology and Medicine, Brown University, Providence, Rhode Island.

Morgan, M. Granger, University Professor, Lord Chair Professor of Engineering, Head, Department of Engineering and Public Policy, Professor of Engineering and Public Policy and of Electrical and Computer Engineering, and Professor, The H. John Heinz III School of Public Policy and Management, Carnegie Mellon University, Pittsburgh, Pennsylvania.

Parson, Edward A. "Ted," Joseph L. Sax Collegiate Professor of Law and Professor of Natural Resources and Environment, University of Michigan, Ann Arbor, Michigan; Senior Research Associate, Centre for Global Studies, University of Victoria, British Columbia, Canada.

Rasch, Philip, Chief Scientist for Climate Science and Laboratory Fellow, Pacific Northwest National Laboratory, Richland, Washington.

Ravishankara, A. R., Director, Chemical Sciences Division, Earth System Research Laboratory, National Oceanic and Atmospheric Administration, Boulder, Colorado; Assistant Professor, Department of Chemistry and Biochemistry, and Affiliate, Cooperative Institute for Research in Environmental Sciences, University of Colorado, Boulder, Colorado.

Robock, Alan, Distinguished Professor (Professor II), Department of Environmental Sciences; Associate Director, Center for Environmental Prediction; Director, Meteorology Undergraduate Program; Member, Graduate Program in Atmospheric Science, Rutgers University, New Brunswick, New Jersey.

Rothstein, Lewis M., Professor of Oceanography, Graduate School of Oceanography and Treasurer, Metcalf Institute Advisory Board, Metcalf Institute for Marine and Environmental Reporting, University of Rhode Island, Narragansett, Rhode Island.

Schneider, Stephen H., Melvin and Joan Lane Professor for Interdisciplinary Environmental Studies and Professor, Department of Biology, Stanford University; Senior Fellow, Woods Institute for the Environment; Professor, by courtesy, Civil and Environmental Engineering, Stanford University, Stanford, California (deceased July 19, 2010).

Shepherd, John, Professorial Research Fellow in Earth System Science, School of Ocean and Earth Science, National Oceanography Centre, University of Southampton, Southampton, United Kingdom.

Socolow, Robert H., Professor, Department of Aerospace and Mechanical Engineering; Co-Director, Carbon Mitigation Initiative, Princeton University, Princeton, New Jersey.

Somerville, Richard C. J., Distinguished Professor Emeritus and Research Professor, Scripps Institution of Oceanography, University of California at San Diego, La Jolla; Team Member, National Science Foundation Science and Technology Center for Multiscaling Modeling of Atmospheric Processes, Department of Atmospheric Science, Colorado State University, Fort Collins, Colorado.

Strand, Stuart E., Research Professor, Department of Civil and Environmental Engineering and School of Forest Resources, University of Washington, Seattle, Washington.

Tilmes, Simone, Project Scientist I, Middle-Upper Atmosphere and WACCM Group, The Earth and Sun Systems Laboratory, National Center for Atmospheric Research, Boulder, Colorado.

Tombari, John, President, Schlumberger Carbon Services, Houston, Texas.

8.3 Foresight scenarios

This appendix contains the four scenarios depicting alternative futures that six external experts worked with us to develop for use in this technology assessment. Our purposes in developing these scenarios included (1) illustrating how some experts view alternative possible futures (2010 to 2030) and judge resulting risk levels (for 2030 and later years) and

(2) stimulating other experts' thinking about the future and eliciting their views.[94]

Developing these scenarios constituted the first of three steps we took to elicit a range of views about the future. In the second and third steps, we asked other experts to express views or comment in response to the scenarios. Specifically,

- in step 2, 28 experts responded to the scenarios; and

- in step 3, 11 additional experts responded to a description of the scenarios and a summary that synthesized step-2 comments and views about the future.

Although some commenters at both steps critiqued or suggested improvements to the scenarios (on points that concern, for example, the effect of carbon constraints, the dollar values associated with carbon constraints or research, and the specification of risk), this appendix presents the scenarios as they were when the 28 commenters first saw them.[95] Our report's methodology is detailed in section 8.1; the range of views experts expressed across our three-step process is represented in the body of the report.

It is important to keep in mind several characteristics of the four scenarios. First, one expert reviewing the scenarios drew our attention to a two-part explanation of how carbon constraints could affect CDR research (which the scenarios do not describe): (1) establishing

---

[94] The six external experts who participated in building the scenarios are listed in section 8.4. Additionally, GAO's Chief Scientist, Timothy Persons, served as host and ex officio member of the group.

[95] The only exception consists of minor corrections to a footnote.

and maintaining a federal research program that includes a significant CDR component is more likely when people are confident that CDR technologies will be used once successfully developed and (2) establishing carbon constraints could encourage the expectation that investing in CDR research is worthwhile.

Next, whereas two of the scenarios (II and IV) specify a degree of carbon constraint that is equivalent to the effect of an international price on CO_2 emissions applicable across all sectors of all major emitting nations, the effect of such a price is not comparable to the effect of prices established for limited sectors or regions.

Further, Scenario II assumes "modest" research funding starting at "tens of millions of dollars" for a program involving several agencies. This assumption was intended to apply to a dedicated research effort for climate engineering that excluded large-scale testing and deployment of any of the technologies (which would be much more expensive). It was not intended to include relevant but separate research in a variety of federal agencies. (Scenarios III and IV, which describe greater research efforts, do not specify funding levels. We discuss uncertainties about funding in the body of the report.)

Finally, all four scenarios give examples of risk for 2030 and later years. They present judgments about levels of risk across three potential developments—a future climate emergency and response (that could involve decision risks), continued future warming (that could be associated with risks from climate change), and no future warming (that might possibly be associated in some scenarios with having risked resources to prepare for a threat that did not occur). Risk levels vary across the scenarios and represent the combined effects of factors that are varied across the scenarios, including

different levels of (1) climate engineering research 2010–30, and the technologies and information developed from it, and (2) other factors in the scenarios such as emissions reduction.

We present two key caveats concerning risk levels. The scenarios present risk levels that represent (1) inexact qualitative judgments that may account for probability and potential severity and (2) judgments about degree of risk that are not necessarily comparable across the three potential developments.[96] Nevertheless, comparisons can be made across scenarios. For example, high decision risk in one scenario and medium decision risk in another implies a judgment that decision risk is higher in one than in the other.

Finally, we note that the scenarios' diagrams of risk levels and CO_2 concentrations are not exact but are, instead, illustrative approximations.

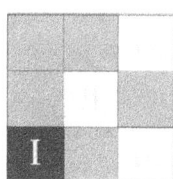 **Scenario I: Status quo**

Between 2010 and 2015, various efforts to jumpstart global agreement on carbon constraints have only token success.[97] Subsequent global efforts to stem carbon dioxide (CO_2) and other greenhouse gas emissions also fail. Individual countries that favor reducing emissions realize that they cannot "go it alone" economically. Adaptation and mitigation continue on paths set earlier.

---

[96] MITRE (2011) illustrates how qualitative judgments of probability and severity may be combined according to risk management literature.

[97] The twin goals of these efforts are to accelerate mitigation efforts (that is, reduce carbon dioxide, or CO_2, emissions) and raise incentives for private-sector research, including research on carbon dioxide removal (CDR) and sequestration.

Americans have diverse views about climate change, and those who are aware of geoengineering approaches remain skeptical about their safety and utility. Debates on global carbon constraints and U.S. geoengineering research programs are limited to a small community of academics, interest groups, and national decision-makers. Proposed federal legislation to establish a research program states four main goals: (1) develop inexpensive, scalable carbon dioxide removal (CDR) and sequestration methods (using mechanical or biological approaches); (2) understand and evaluate fast-acting methods like stratospheric aerosol injection, including the modeling of potential side effects and total cost analyses; (3) involve other nations' governments and scientists in joint research and in setting international research guidelines and limitations; and (4) inform decision-makers about systemic risks and tradeoffs among various geoengineering technologies and between these and other climate change approaches. But congressional efforts to enact legislation fail, despite the support of nearly half the Congress.

Without carbon constraints to stimulate private-sector research and development (R&D) and without a federal research program dedicated to geoengineering, U.S. scientists focus their efforts on other areas. The United States makes rapid advances in emerging areas such as synthetic biology and nanotechnology, but applications to geoengineering are limited. Various other nations (and some private sector organizations) develop fast-acting technologies for use in a climate emergency, but they do not always focus on identifying side effects or share their results with the global scientific community. Efforts to develop international guidelines that limit field tests and deployment fail.

Risks across three potential developments (2030 and later years):

- *Climate emergency and response:* Immediate decision risks would be very high if a sudden acceleration of the warming trend occurred spontaneously. World leaders would be under pressure to make decisions quickly—and might opt to use fast-acting, risky geoengineering technologies—despite inadequate information on their effectiveness and side effects.[98] (See red bar in illustration.)

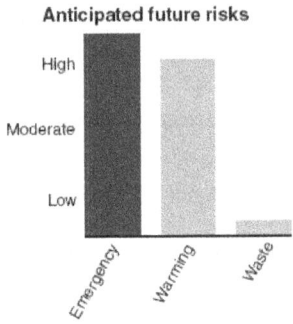

Anticipated future risks

- *Continued warming:* Future global climate risks would be high. As of 2030, this scenario sets the United States and, indeed, the world, on a path of increasing CO_2 emissions and rising atmospheric concentrations. Decades of CDR research starting in 2030 would be needed prior to deployment aimed at decreasing future CO_2 build-up. Prospects thus include temperature increase and far future sea level rise that might engulf vulnerable areas, naval installations, and so forth; such a future might also bring other very serious negative consequences on a global scale. (See the line chart and the orange bar in the bar chart.)

[98] Similarly, without adequate information on fast-acting technologies, it would be difficult for leaders to decide how to respond to a surprise deployment by a single nation, terrorist group, or some other "rogue" geoengineering effort.

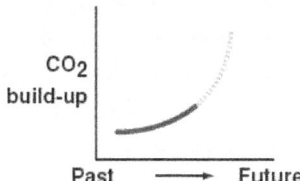

CO₂
build-up

Past ———→ Future

- *Resources wasted on geoengineering, if no future warming:* Risks of having wasted efforts and expenditures would be near zero. This scenario commits few, if any, new resources. (See gray bar in the bar chart.)

Scenario II: Some action

By 2015 or soon thereafter, somewhat improved data and models of climate change reduce uncertainty and appear to validate earlier conclusions about global warming caused by human activity. A series of extreme weather developments causes widespread concern. As a result, major emitting nations agree to new carbon constraints with strong enforcement, but the reduction goals and guidelines are limited (equivalent to a $10 to $15 price on a ton of CO_2). These measures slightly increase both mitigation efforts and existing incentives for private-sector R&D on direct air capture and sequestration of CO_2. Scientists expect these changes will not stabilize future accumulated levels of CO_2 but may delay a far-off climate emergency by about 10 years and represent a start.[99] Additionally, U.S. legislation

establishes a modest geoengineering research program that involves several federal agencies.[100] The funding level is tens of millions of dollars the first year, with plans for modest annual increases. Public acceptance of these developments is mixed. The modest research program has no public engagement or outreach component.

Without adequate information on the general public's views and concerns about geoengineering, the government and scientists do not craft the research program in a way that encourages public acceptance, and inadvertently they alienate some original supporters. Some years of benign weather intervene. While the research program continues to receive its original level of support, the planned annual expenditure increases are not put into effect. The research program leverages its limited funds by encouraging the private sector to develop new methods of direct CO_2 air capture and sequestration that are somewhat less expensive than technologies developed through 2010. But both in the United States and around the globe, industries that emit significant CO_2 are not eager to purchase the new technologies to offset emissions: the limited carbon constraints have not created a sufficient incentive. The research program also makes some advances in developing and evaluating fast-acting methods like stratospheric aerosol injection, but research by others outpaces the federal effort. Some new fast-acting, high-impact technologies are not rigorously evaluated for side effects. Results are not always shared with the global scientific community. Thus, we lack key information on some new methods and their implications.

---

[99] The moderate reductions are not at the scale required to transform energy or energy-intensive industrial sectors.

[100] One option, among others, for housing a dedicated research program, would be the U.S. Global Climate Change Research Program.

Risks across three potential developments (2030 and later years):

- *Climate emergency and response:* Immediate decision risks would be moderate to high. By 2030, world leaders responding to an emergency would have some geoengineering information to guide them, but the information would be inadequate for some new technologies.[101]

Anticipated future risks

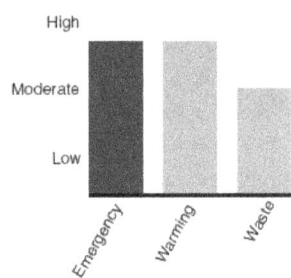

- *Continued warming:* Future global climate risks would be moderate to high. As of 2030, more is known about carbon emissions and controls than in Scenario I. Still, starting serious R&D on CDR in 2030 would mean years or decades of delay before deployment. The world would likely be on a path of continued build-up of CO_2 concentrations—although its trajectory would be slightly slower/lower than in Scenario I. The prospect of negative consequences like sea level rise would still loom, eventually, in the far future.

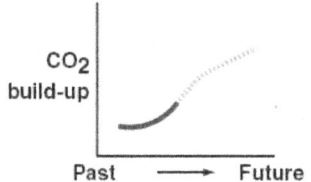

- *Resources wasted on geoengineering if no future warming:* Risks of having wasted efforts and expenditures would be moderate. In the absence of warming, some new geoengineering technologies would not be useful, but others might serve other purposes, such as helping to reduce ocean acidification.

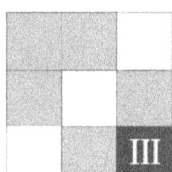

Scenario III: Action on research but not carbon

By 2015 or soon thereafter, significantly improved data and models of climate change appear to validate earlier conclusions about anthropogenic global warming. Highly disruptive and extreme weather events affect the United States and many other nations, causing waves of concern and even, periodically, a crisis atmosphere. Other nations pursue geoengineering research, a fact that is widely reported. The balance of U.S. public opinion turns toward taking action on climate change, despite opposition from some at home and the lack of global agreement on carbon constraints.

Although public opinion generally favors climate action, some opinion leaders believe that, economically, the United States cannot

---

[101] Also, relative to Scenario I, the increased knowledge might better prepare decision-makers for responding to a "rogue" deployment.

"go it alone" in legislating carbon constraints. Those who are opposed emphasize this point, and legislative measures to step up U.S. emission controls fail on a close vote. At the same time, the Congress and the president work together successfully to design and build support for legislation that establishes an aggressive federal geoengineering research program, starting with moderate resources but progressing toward a major funding commitment. The research program involves public engagement to build support in the years ahead (including years in which extreme climate events may not occur); establishes an adaptive strategy that entails periodic reviews by an external body such as the National Academies and horizon scans to identify new opportunities; promotes innovation through creative incentives, such as federal contests with cash awards, in addition to using more conventional approaches; and emphasizes international cooperation. The main goals of this research program are similar to those in the failed legislation outlined in Scenario I (points 1–4).

As a result, major advances are made in developing, understanding, and evaluating fast-acting methods (like next-generation stratospheric aerosol and injection methods); understanding tradeoffs among different approaches; building new approaches that reduce the potential for side effects; and furthering basic science concerning climate change. Other advances are made in international cooperation on research limitations and guidelines for the use of geoengineering. Additionally, the research program helps develop potentially transformative methods of direct CO_2 air capture and sequestration. These new technologies cost substantially less than 2010 technologies but, given the lack of carbon constraints, there are virtually no incentives for emitting industries to buy them. These technologies often fall into the "valley of

death" between R&D and commercial success and large-scale deployment. Researchers and commercial firms become discouraged. The focus on direct air capture and sequestration suffers some loss of credibility (that is, the government is seen as investing in unused technologies), and it is significantly cut back.

Risks across three potential developments (2030 and later years):

• *Climate emergency and response:* Immediate decision risks would be moderate. By 2030, decision-makers have information to support decisions about the use (or nonuse) of fast-acting geoengineering technologies. Catastrophic risks are minimized.[102]

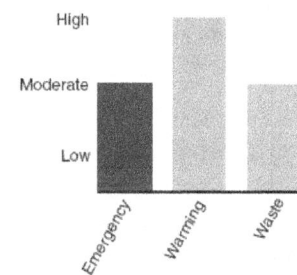

Anticipated future risks

• *Continued warming:* Future global climate risks would be high. Knowledge has increased somewhat but—without utilization of CDR—the world is still likely on a path of building up the concentration of CO_2 in the atmosphere. This brings the prospect of higher temperatures

---

[102] Note, however, that in this scenario, decision-makers who reject fast-acting technologies would lack alternative, more gradual approaches for dealing with the problem. For example, because CDR technologies were "left on the drawing board" rather than being further developed and deployed, decision-makers would not have the option of ramping up existing direct air capture efforts. Decades would be likely to be needed to prepare for such an effort.

which imply, in the far future, a sea level rise and the possible consequences in Scenario I.

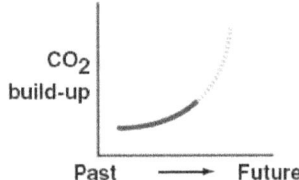

- *Resources wasted on geoengineering if no future warming:* Risks of having wasted efforts and expenditures would be moderate. The financial losses and efforts in a federal research program designed specifically to combat warming could be somewhat offset if some new technologies can be used to address ocean acidification or to develop spin-off technologies to apply in other areas.

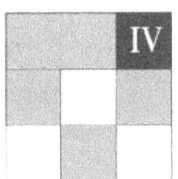 **IV** Scenario IV: Major action

By 2015 or soon thereafter, significantly stronger climate-change data and models will have reduced uncertainty, deepened understanding, and validated earlier scientific conclusions. Also during this half decade, several unprecedented, highly disruptive, and extreme weather events will affect a number of nations (including the United States), causing mass deaths, migration, and devastating property damage. In a jarring development, one nation unilaterally stages a major real-world test of a fast-acting geoengineering technology in a remote area—without first warning other nations. The test's negative effects are limited, but there is a step jump increase in global recognition of the need for coordination and cooperation.

In the United States, the balance of public opinion tips toward favoring an aggressive lowering of climate risks. Taking a leadership role, U.S. envoys help achieve a global agreement on relatively aggressive carbon constraints (equivalent to a carbon price of $30 per ton of CO_2). The global carbon constraints create a worldwide incentive for the private sector to pursue mitigation strategies, such as alternative fuels and renewables, as well as geoengineering approaches like scalable, direct air capture and sequestration. A new presidential-congressional initiative establishes an aggressive, innovative, and adaptive geoengineering research program that cuts across multiple agencies. It includes strong international cooperation and other goals similar to those in the failed legislation outlined in Scenario I (points 1–4). Additionally, this initiative emphasizes adaptation, research innovation, and public engagement.

In part because of this research program, new developments in areas such as synthetic biology and nanotechnology are applied to geoengineering (and to other areas such as energy production and conservation), resulting in a number of potentially game-changing breakthroughs. The new U.S. initiative sets in motion a range of programs and policies to ensure that new technologies will have opportunities to (1) transform energy sectors and help lower future emissions in the United States and around the globe, (2) reduce existing and continuing build-up of CO_2 through air capture (because emissions reduction will not be complete), and (3) improve the U.S. economic and export profile. Measures to spur dissemination of new technologies include, for example, working with states and regions to develop targeted sector or regional plans, as well as international coordination.

Additionally, the research program includes evaluations of side effects; analyses of economic, legal, and social implications; and analyses of tradeoffs and systemic risk—to help inform policymakers and the interested public. Overall, the program's public engagement feature and its effectuation of economic gains and international cooperation help sustain support for this initiative through 2030.

Risks across three potential developments (2030 and later years):

• *Climate emergency/response:* Immediate decision risks are low to moderate. By 2030, U.S. decision-makers and the global community would have information that helps prepare them for responding to a climate emergency. Additionally, there would be international mechanisms in place to support global cooperation, and thus help avoid conflicts.

Anticipated future risks

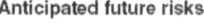
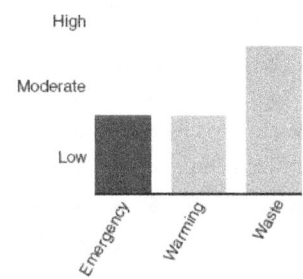

• *Continued warming:* Future global climate risks would be low to moderate. By 2030, the world is on a path toward eventual stabilization and subsequent reduction of CO_2 build-up—hence, less warming. Although some sea level rise may occur, the overall prospects for negative

consequences in the far future would be substantially reduced relative to Scenarios I–III.

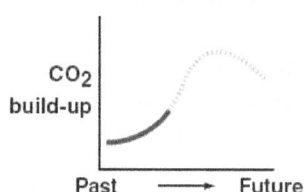

• *Resources wasted on geoengineering if no future warming:* Risks would be moderate to high. In this scenario, very large investments (in terms of both financial resources and efforts that might have been used in other ways) would have been made, and unrecoverable losses could be significant. As in Scenarios II and III, if discoveries and technologies developed as a result of geoengineering research were able to be used in other ways, losses could be mitigated—for example, by helping to reduce ocean acidification. In the longer term, some of the geoengineering technologies developed to combat warming might be used instead to help avoid other adverse affects that might be associated with extremely high concentrations of CO_2.

8.4 The six external experts who participated in building the scenarios

Cannizzaro, Christopher, Physical Science Officer / AAAS Science and Technology Policy Fellow, Office of Space and Advanced Technology (OES/SAT), U.S. Department of State, Washington, D.C.

Gallaudet, **Tim,** Deputy Director, Navy's Task Force Climate Change, Office of the Oceanographer of the Navy, Chief of Naval Operations Staff, Washington, D.C.

Lackner, **Klaus,** Department Chair, Ewing and J. Lamar Worzel Professor of Geophysics, Earth and Environmental Engineering; Director, Lenfest Center for Sustainable Energy, The Earth Institute, Columbia University, New York.

Patrinos, **Aristides A. N.,** President, Synthetic Genomics, La Jolla, California and Washington, D.C.

Rasch, **Philip,** Chief Scientist for Climate Science and Laboratory Fellow, Pacific Northwest National Laboratory, Richland, Washington.

Rejeski, **David,** Director, Science and Technology Innovation Program, and Director, Project on Emerging Nanotechnologies, Woodrow Wilson International Center for Scholars, Washington, D.C.

8.5 Experts who commented in response to the scenarios

Barrett, **Scott,** Lenfest-Earth Institute Professor of Natural Resource Economics, School of International and Public Affairs and Earth Institute, Columbia University, New York.

Beck, **Robert A.,** Executive Vice President and Chief Operating Officer, National Coal Council Inc., Washington, D.C.

Bronson, **Diana,** Programme Manager and Researcher, ETC Group, Ottawa, Ontario, Canada.

Bunzl, **Martin,** Professor, Department of Philosophy, Rutgers University; Director, Rutgers

Initiative on Climate and Social Policy, Rutgers University, New Brunswick, New Jersey.

Carlson, **Rob,** Principal, Biodesic, Seattle, Washington.

Cascio, **Jamais,** Senior Fellow, Institute for Emerging Ethics and Technologies, Hartford, Connecticut; Director of Impacts Analysis, Center for Responsible Nanotechnology, Menlo Park, California; Research Fellow, Institute for the Future, Palo Alto, California.

Christy, **John R.,** Distinguished Professor of Atmospheric Science and Director, Earth System Science Center, University of Alabama, Huntsville, Alabama; Alabama State Climatologist, The Alabama Office of the State Climatologist, Huntsville, Alabama.

Fetter, **Steve,** Assistant Director At-Large, U.S. Office of Science and Technology Policy, Executive Office of the President of the United States, Washington, D.C.

Fleming, **James R.,** Professor and Director of Science, Technology, and Society Program, Colby College, Waterville, Maine.

Hamilton, **Clive,** Professor of Public Ethics, Centre for Applied Philosophy and Public Ethics, a joint center of the Australian National University, Charles Sturt University, and the University of Melbourne, Melbourne, Australia; Vice-Chancellor's Chair, Charles Sturt University, Sydney, Australia.

Hawkins, **David G.,** Director of Climate Programs, Natural Resources Defense Council, New York, New York.

Hsing, **Helen,** Managing Director, Strategic Planning and External Liaison, U.S. Government Accountability Office, Washington, D.C.

Johnson, Jean, Executive Vice President, Director of Education Insights and Director of Programs, Public Agenda, New York, New York.

Khosla, Vinod, Partner, Khosla Ventures, Menlo Park, California.

Lane, Lee, Visiting Fellow, Hudson Institute, Washington, D.C.

Lomborg, Bjørn, Director, Copenhagen Consensus Center, Denmark.

Long, Jane C. S., Associate Director, Energy and Environment Directorate, Lawrence Livermore National Laboratory, Livermore, California.

MacCracken, Michael, Chief Scientist for Climate Change Programs, Climate Institute, Washington, D.C.

Maynard, Andrew D., Director, University of Michigan Risk Science Center, and Professor, Environmental Health Sciences, School of Public Health, University of Michigan, Ann Arbor, Michigan.

Olson, Robert L., Senior Fellow, Institute for Alternative Futures, Alexandria, Virginia.

Robock, Alan, Distinguished Professor (Professor II), Department of Environmental Sciences; Associate Director, Center for Environmental Prediction; Director, Meteorology Undergraduate Program; Member, Graduate Program in Atmospheric Science, Rutgers University, New Brunswick, New Jersey.

Sarewitz, Daniel, Co-Director, Consortium for Science, Policy & Outcomes; Associate Director, Center for Nanotechnology in Society; Professor of Science and Society, College of Liberal Arts and Sciences; and Professor, School of Life

Sciences and School of Sustainability, Arizona State University, Tempe, Arizona.

Schneider, John P., Deputy Director for Research, Earth System Research Laboratory, National Oceanic and Atmospheric Administration, Boulder, Colorado.

Seidel, Stephen, Vice President for Policy Analysis and General Counsel, Pew Center on Global Climate Change, Arlington, Virginia.

Suarez, Pablo, Associate Director of Programmes, Red Cross/Red Crescent Climate Centre, The Hague, The Netherlands; Visiting Fellow, Frederick S. Pardee Center for the Study of the Longer-Range Future, Boston University, Boston, Massachusetts.

Victor, David G., Professor, International School of International Relations and Pacific Studies; Director, Laboratory on International Law and Regulation, University of California at San Diego, San Diego.

Wiener, Jonathan B., William R. and Thomas L. Perkins Professor of Law and Director, JD-LLM Program in International and Comparative Law, Duke Law School, Durham, North Carolina; Professor of Environmental Policy, Nicholas School of the Environment and Professor of Public Policy, Sanford School of Public Policy, Duke University, Durham, N.C.; Fellow, Resources for the Future, Wash., D.C.

Wilcoxen, Peter J., Director, Center for Environmental Policy and Administration, and Associate Professor, Economics and Public Administration, The Maxwell School, Syracuse University, Syracuse, New York; Nonresident Senior Fellow, Economic Studies, The Brookings Institution, Washington, D.C.

8.6 Experts who participated in our meeting on climate engineering

The 11 of these experts whose names are starred (*) both (1) commented on the future of climate engineering during the Meeting and (2) had not previously participated in either constructing the scenarios or commenting on them. We discussed the views of these experts in section 4 of this report (Experts' Views of the Future of Climate Engineering Research). John Latham was scheduled to attend this meeting but was unable to be there and provided written comments instead.

*Berg, Robert J., Trustee, World Academy of Art and Science, Pittsburgh, Pennsylvania; Senior Advisor, World Federation of United Nations Associations, New York, New York.

Bunzl, Martin, Professor, Department of Philosophy, Rutgers University; Director, Rutgers Initiative on Climate and Social Policy, Rutgers University, New Brunswick, New Jersey.

*Duren, Riley, Chief Systems Engineer, Earth Science and Technology Directorate, Jet Propulsion Laboratory, California Institute of Technology, Pasadena, California.

Espinal, Laura, Materials Scientist, Ceramics Division, Functional Properties Group, Material Measurement Laboratory, National Institute of Standards and Technology, Gaithersburg, Maryland.

Fetter, Steve, Assistant Director At-Large, U.S. Office of Science and Technology Policy, Executive Office of the President of the United States, Washington, D.C.

*Fraser, Gerald T., Chief, Optical Technology Division, Physical Measurement Laboratory, National Institute of Standards and Technology, Gaithersburg, Maryland.

*Hunter, Kenneth W., Senior Fellow, Institute for Global Chinese Affairs; Senior Fellow, Joint Institute for Food Safety and Applied Nutrition, University of Maryland, College Park, Maryland.

*Janetos, Anthony, Director, Joint Global Change Research Institute, Pacific Northwest National Laboratory and University of Maryland, College Park, Maryland.

Johnson, Jean, Executive Vice President, Director of Education Insights and Director of Programs, Public Agenda, New York, New York.

*LaPorte, Todd R., Professor Emeritus of Political Science, University of California, Berkeley, California.

*Lehmann, Christopher Johannes, Associate Professor, Department of Crop and Soil Sciences, Cornell University, Ithaca, New York.

Long, Jane C. S., Associate Director, Energy and Environment Directorate, Lawrence Livermore National Laboratory, Livermore, California.

MacCracken, Michael, Chief Scientist for Climate Change Programs, Climate Institute, Washington, D.C.

*MacDonald, Alexander E. "Sandy," Director, Earth System Research Laboratory; and Deputy Assistant Administrator for Laboratories and Cooperative Institutes, Office of Oceanic and Atmospheric Research, National Oceanic and Atmospheric Administration, Boulder, Colorado.

Rejeski, David, Director, Science and Technology Innovation Program, and Director, Project on Emerging Nanotechnologies, Woodrow Wilson International Center for Scholars, Washington, D.C.

*****St. John, Courtney,** Climate Change Affairs Officer, Navy's Task Force Climate Change, Office of the Oceanographer of the Navy, Chief of Naval Operations Staff, Washington, D.C.

Thernstrom, Samuel, Senior Policy Advisor, Geoengineering Task Force, Bipartisan Policy Center, Washington, D.C.; Senior Climate Policy Advisor, Clean Air Task Force, Boston, Massachusetts.

*****Tilmes, Simone,** Project Scientist I, Chemistry Climate Liaison, Atmospheric Chemistry Division, National Center for Atmospheric Research, Boulder, Colorado.

Tombari, John, President, Schlumberger Carbon Services, Houston, Texas.

*****Toon, Owen B.,** Chair and Professor, Department of Atmospheric and Oceanic Sciences; Research Associate, Laboratory for Atmospheric and Space Physics; Director, Toon Aerosol Research Group, University of Colorado, Boulder.

8.7 Reviewers of the report draft

Duren, Riley, Chief Systems Engineer, Earth Science and Technology Directorate, Jet Propulsion Laboratory, California Institute of Technology, Pasadena, California.

Espinal, Laura, Materials Scientist, Ceramics Division, Functional Properties Group,

Material Measurement Laboratory, National Institute of Standards and Technology, Gaithersburg, Maryland.

Fraser, Gerald T., Chief, Optical Technology Division, Physical Measurement Laboratory, National Institute of Standards and Technology, Gaithersburg, Maryland.

Hunter, Kenneth W., Senior Fellow, Institute for Global Chinese Affairs; Senior Fellow, Joint Institute for Food Safety and Applied Nutrition, University of Maryland, College Park, Maryland.

Janetos, Anthony, Director, Joint Global Change Research Institute, Pacific Northwest National Laboratory, and University of Maryland, College Park, Maryland.

Johnson, Jean, Executive Vice President, Director of Education Insights and Director of Programs, Public Agenda, New York, New York.

LaPorte, Todd R., Professor Emeritus of Political Science, University of California, Berkeley.

Latham, John, Emeritus Professor of Physics, University of Manchester, UK; Visiting Professor, University of Leeds, UK; Senior Research Associate, National Center for Atmospheric Research, Boulder, Colorado.

Lehmann, Christopher Johannes, Associate Professor, Department of Crop and Soil Sciences, Cornell University, Ithaca, New York.

MacCracken, Michael, Chief Scientist for Climate Change Programs, Climate Institute, Washington, D.C.

MacDonald, Alexander E. "Sandy," Director, Earth System Research Laboratory, and Deputy

Assistant Administrator for Laboratories and Cooperative Institutes, Office of Oceanic and Atmospheric Research, National Oceanic and Atmospheric Administration, Boulder, Colorado.

St. John, Courtney, Climate Change Affairs Officer, Navy's Task Force Climate Change, Office of the Oceanographer of the Navy, Chief of Naval Operations Staff, Washington, D.C.

Thernstrom, Samuel, Senior Policy Advisor, Geoengineering Task Force, Bipartisan Policy Center, Washington, D.C.; Senior Climate Policy Advisor, Clean Air Task Force, Boston, Massachusetts.

Tilmes, Simone, Project Scientist I, Chemistry Climate Liaison, Atmospheric Chemistry Division, National Center for Atmospheric Research, Boulder, Colorado.

Toon, Owen B., Chair and Professor, Department of Atmospheric and Oceanic Sciences; Research Associate, Laboratory for Atmospheric and Space Physics; Director, Toon Aerosol Research Group, University of Colorado, Boulder.

9 References

Akbari, Hashem, Surabi Menon, and Arthur Rosenfeld. 2009. "Global Cooling: Increasing World-Wide Urban Albedos to Offset CO_2." *Climatic Change* 94 (3–4): 275–86.

Angel, Roger. 2006. "Feasibility of Cooling the Earth with a Cloud of Small Spacecraft Near the Inner Lagrange Point (L1)." *Proceedings of the National Academy of Sciences of the USA* 103 (46): 17184–89.

Asilomar Scientific Organizing Committee. 2010. *The Asilomar Conference Recommendations on Principles for Research into Climate Engineering Techniques: Conference Report.* Washington, D.C.: Climate Institute, November.

Azar, Christian, Kristian Lindgren, Eric Larson, and Kenneth Möllersten. 2006. "Carbon Capture and Storage from Fossil Fuels and Biomass— Costs and Potential Role in Stabilizing the Atmosphere." *Climatic Change* 74 (1–3): 47–79.

Backlund, P., A. Janetos, D. Schimel, J. Hatfield, K. Boote, P. Fay, L. Hahn, C. Izaurralde, B. A. Kimball, T. Mader, J. Morgan, D. Ort, W. Polley, A. Thomson, D. Wolfe, M. G. Ryan, S. R. Archer, R. Birdsey, C. Dahm, L. Heath, J. Hicke, D. Hollinger, T. Huxman, G. Okin, R. Oren, J. Randerson, W. Schlesinger, D. Lettenmaier, D. Major, L. Poff, S. Running, L. Hansen, D. Inouye, B. P. Kelly, L. Meyerson, B. Peterson, and R. Shaw. 2008. *The Effects of Climate Change on Agriculture, Land Resources, Water Resources, and Biodiversity in the United States.* Report by the U.S. Climate Change Science Program and the Subcommittee on Global Change Research. Climate Synthesis and Assessment Product 4.3. Washington, D.C.: U.S. Department of Agriculture, May.

Bader, David C., Curt Covey, William J. Gutowski Jr., Isaac M. Held, Kenneth E. Kunkel, Ronald L. Miller, Robin T. Tokmakian, and Minghua H. Zhang. 2008. *Climate Models: An Assessment of Strengths and Limitations.* Synthesis and Assessment Product 3.1. Report by the U.S. Climate Change Science Program and the Subcommittee on Global Change Research. Washington, D.C.: U.S. Department of Energy, Office of Biological and Environmental Research, July.

Baird, Colin. 1998. *Environmental Chemistry.* 2nd ed. New York: W. H. Freeman and Company.

Bala, G., Ken Caldeira, Rama Nemani, Long Cao, George Ban-Weiss, and Ho-Jeong Shin. 2010. "Albedo Enhancement of Marine Clouds to Counteract Global Warming: Impacts on the Hydrological Cycle." *Climate Dynamics, Online First™*, June 23. www.springerlink.com/content/9569172415150486.

Barrett, Scott. 2008. "The Incredible Economics of Geoengineering." *Environmental and Resource Economics* 39 (1): 45–54.

Barrett, Scott. 1998. "Political Economy of the Kyoto Protocol." *Oxford Review of Economic Policy* 14 (4): 20–39.

Bertram, Christine. 2009. "Ocean Iron Fertilization: An Option for Mitigating Climate Change?" Kiel Policy Brief No. 3. Kiel, Germany: Kiel Institute for the World Economy, March.

Bittle, Scott, Jonathan Rochkind, and Amber Ott. 2009. *The Energy Learning Curve: Coming from Different Starting Points, the Public Sees Similar Solutions.* New York: Public Agenda.

Boden, T. A., G. Marland, and R. J. Andres. 2010. *Global, Regional, and National Fossil-Fuel CO_2 Emissions.* Carbon Dioxide Information Analysis Center, Oak Ridge National Laboratory. Oak Ridge, Tenn.: U.S. Department of Energy.

Boyd, Philip W. 2008. "Introduction and Synthesis" (Theme Section: Implications of Large-Scale Iron Fertilization of the Oceans). *Marine Ecology Progress Series* 364: 213–18, July 29.

Buesseler, Ken O., Scott C. Doney, David M. Karl, Philip W. Boyd, Ken Caldeira, Fei Chai, Kenneth H. Coale, Hein J. W. de Baar, Paul G. Falkowski, Kenneth S. Johnson, Richard S. Lampitt, Anthony F. Michaels, S. W. A. Naqvi, Victor Smetacek, Shigenobu Takeda, and Andrew J. Watson. 2008a. "Ocean Iron Fertilization—Moving Forward in a Sea of Uncertainty." *Science* 319 (5860): 161.

Buesseler, Ken, Scott Doney, and Hauke Kite-Powell, eds. 2008b. "Should We Fertilize the Ocean to Reduce Greenhouse Gases?" *Oceanus* 46 (1).

Caldeira, Ken, and Lowell Wood. 2008. "Global and Arctic Climate Engineering: Numerical Model Studies." *Philosophical Transactions of the Royal Society A* 366: 4039–56.

Canadell, Josep G., and Michael R. Raupach. 2008. "Managing Forests for Climate Change Mitigation." *Science* 320 (5882): 1456–57.

Carbo, Michiel C., Ruben Smith, Bram van der Drift, and Daniel Jansen. 2010. "Opportunities for BioSNG Production with CCS." 1st International Workshop on Biomass & CCS, Orléans, France, October 14–15.

CMI (Carbon Mitigation Initiative). 2010. *Carbon Mitigation Initiative: Annual Report 2009.* Princeton, N.J.: Princeton University.

CNA Corporation. 2007. *National Security and the Threat of Climate Change.* Alexandria, Va.: The CNA Corporation.

DOE (U.S. Department of Energy). 2006. *U.S. Climate Change Technology Program Strategic Plan.* DOE/PI-0005. Washington, D.C.: U.S. Climate Change Technology Program, September.

Dooley, J. J., and C. L. Davidson. 2010. *A Brief Technical Critique of Ehlig-Economides and Economides 2010 "Sequestering Carbon Dioxide in a Closed Underground Volume."* PNNL 19249. Richland, Wash.: Pacific Northwest National Laboratory, April.

Dyer, Gwynne. 2010. *Climate Wars: The Fight for Survival as the World Overheats.* Oxford, Eng.: Oneworld Publications.

Early, James T. 1989. "Space-Based Solar Shield to Offset Greenhouse Effect." *Journal of the British Interplanetary Society* 42: 567–69.

Ehlig-Economides, Christine, and Michael J. Economides. 2010. "Sequestering Carbon Dioxide in a Closed Underground Volume." *Journal of Petroleum Science and Engineering* 70 (1–2): 123–30.

Eilperin, Juliet. 2010. "Maldives' Fight on Carbon Status Is One of Survival." *Washington Post,* October 11, p. A5.

Field, Christopher B., Jorge Sarmiento, and Burke Hales. 2007. "The Carbon Cycle of North

America in a Global Context." In *The First State of the Carbon Cycle Report (SOCCR): The North American Carbon Budget and Implications for the Global Carbon Cycle,* Anthony W. King, Lisa Dilling, Gregory P. Zimmerman, David M. Fairman, Richard A. Houghton, Gregg Marland, Adam Z. Rose, and Thomas J. Wilbanks, eds., pp. 21–28. Synthesis and Assessment Product 2.2. Report by the U.S. Climate Change Science Program and the Subcommittee on Global Change Research. Asheville, N.C.: NOAA, National Climatic Data Center, November.

Fleming, James Rodger. 2010. *Fixing the Sky: The Checkered History of Weather and Climate Control.* New York: Columbia University Press.

Fraser, G. T., S. W. Brown, R. U. Datla, B. C. Johnson, K. R. Lykke, and J. P. Rice. 2008. "Measurement Science for Climate Remote Sensing." In *Earth Observing Systems XIII,* James J. Butler and Jack Xiong, eds. *Proceedings of SPIE* 7801, 708102 (Proceedings of SPIE Optics and Photonics Conference 2008).

GAO (U.S. Government Accountability Office). 2010a. *Climate Change: A Coordinated Strategy Could Focus Federal Geoengineering Research and Inform Governance Efforts.* GAO-10-903. Washington, D.C.: September 23.

GAO (U.S. Government Accountability Office). 2010b. *Climate Change: Preliminary Observations on Geoengineering Science, Federal Efforts, and Governance Issues.* Testimony before the Committee on Science and Technology, U.S. House of Representatives, 111th Cong. GAO-10-546T. Washington, D.C.: March 18.

GAO (U.S. Government Accountability Office). 2010c. *Coal Power Plants: Opportunities Exist for DOE to Provide Better Information on the Maturity of Key Technologies to Reduce Carbon Dioxide Emissions.* GAO-10-675. Washington, D.C.: June 16.

GAO (U.S. Government Accountability Office). 2009a. *Climate Change Adaptation: Strategic Federal Planning Could Help Government Officials Make More Informed Decisions.* GAO-10-113. Washington, D.C.: October 7.

GAO (U.S. Government Accountability Office). 2009b. *GAO Cost Estimating and Assessment Guide: Best Practices for Developing and Managing Capital Program Costs.* GAO-09-3SP. Washington, D.C.: March.

GAO (U.S. Government Accountability Office). 2009c. *International Space Station: Significant Challenges May Limit Onboard Research.* GAO-10-9. Washington, D.C.: November 25.

GAO (U.S. Government Accountability Office). 2008a. *Climate Change: Expert Opinion on the Economics of Policy Options to Address Climate Change.* GAO-08-605. Washington, D.C.: May 9.

GAO (U.S. Government Accountability Office). 2008b. *Highway Safety: Foresight Issues Challenge DOT's Efforts to Assess and Respond to New Technology-Based Trends.* GAO-09-56. Washington, D.C.: October 3.

GAO (U.S. Government Accountability Office). 1999. *Best Practices: Better Management of Technology Development Can Improve Weapon System Outcomes.* GAO/NSIAD-99-162. Washington, D.C.: July 30.

GAO (U.S. Government Accountability Office). 1995. *Global Warming: Limitations of General Circulation Models and Costs of Modeling Efforts.* GAO/RCED-95-164. Washington, D.C.: July 13.

Gaskill, Alvia. 2004. DOE Meeting Summary: Summary of Meeting with U.S. DOE to Discuss Geoengineering Options to Prevent Abrupt Long-Term Climate Change. Held at the U.S. Department of Energy, Washington, D.C., June 16, 2004. www.global-warming-geo-engineering. org/3/contents.html

Gaskill, Alvia, ed. n.d. *Global Warming Albedo Enhancement.* Sedlescombe East Sussex, Eng.: Library4Science. www.library4science.com (2000–11).

Gordon, Bart. 2010. *Engineering the Climate: Research Needs and Strategies for International Coordination.* Report by the Chairman, Committee on Science and Technology, U.S. House of Representatives, 111th Cong., 2nd Sess. Washington, D.C.: October.

Gornall, Jemma, Richard Betts, Eleanor Burke, Robin Clark, Joanne Camp, Kate Willett, and Andrew Wiltshire. 2010. "Implications of Climate Change for Agricultural Productivity in the Early Twenty-First Century," *Philosophical Transactions of the Royal Society B* 365: 2973–89.

Govindasamy, B., S. Thompson, P. B. Duffy, K. Caldeira, and C. Delire. 2002. "Impact of Geoengineering Schemes on the Terrestrial Biosphere." *Geophysical Research Letters* 29 (22): 2061.

Govindasamy, Bala, and Ken Caldeira. 2000. "Geoengineering Earth's Radiation Balance to Mitigate CO_2-induced Climate Change." *Geophysical Research Letters* 27 (14): 2141– 44.

Graham, John D., and Jonathan Baert Wiener, eds. 1995. *Risk versus Risk: Tradeoffs in Protecting Health and the Environment.* Cambridge, Mass.: Harvard University Press.

Grimstad, Alv-Arne, Sorin Georgescu, Erik Lindeberg, and Jean-Francois Vuillaume. 2009. "Modelling and Simulation of Mechanisms for Leakage of CO_2 from Geological Storage." Proceedings of the 9th International Conference on Greenhouse Gas Control Technologies (GHGT-9), Washington, D.C., November 16–20, 2008. *Energy Procedia* 1 (1): 2511-2518.

Hamwey, Robert M. 2007. "Active Amplification of the Terrestrial Albedo to Mitigate Climate Change: An Exploratory Study." *Mitigation and Adaptation Strategies for Global Change* 12 (4): 419–39.

Hansen, J., R. Ruedy, M. Sato, and K. Lo. 2010. "Global Surface Temperature Change." *Reviews of Geophysics,* 48, RG4004: 1–29.

Hao, Yue, Tom Wolery, and Susan Carroll. 2009. *Preliminary Simulations of CO_2 Transport in the Dolostone Formations in the Ordos Basin, China,* LLNL-TR-412701. Report prepared under the auspices of the U.S. Department of Energy by Lawrence Livermore National Laboratory under Contract DE-AC52-07NA27344. Livermore, Calif.: Lawrence Livermore National Laboratory, May 1.

Heckendorn, P., D. Weisenstein, S. Fueglistaler, B. P. Luo, E. Rozanov, M. Schraner, L. W. Thomason, and T. Peter. 2009. "The Impact of Geoengineering Aerosols on Stratospheric Temperature and Ozone." *Environmental Research Letters* 4 (4):1-12.

Herzog, Howard. 2003. *Assessing the Feasibility of Capturing CO_2 from the Air.* Massachusetts Institute of Technology, Laboratory for Energy and the Environment, Pub.No. MIT LFEE 2003-02 WP. Cambridge, Mass.: Massachusetts Institute of Technology, October.

Houghton, J. T., Y. Ding, D. J. Griggs, M. Noguer, P. J. van der Linden, X. Dai, K. Maskell, and C. A. Johnson, eds. 2001. *Climate Change 2001: The Scientific Basis.* Contribution of Working Group I to the Third Assessment Report of the Intergovernmental Panel on Climate Change. New York and Cambridge, Eng.: Cambridge University Press.

Houghton, R. A., E. A. Davidson, and G. M. Woodwell. 1998. "Missing Sinks, Feedbacks, and Understanding the Role of Terrestrial Ecosystems in the Global Carbon Balance." *Global Biogeochemical Cycles* 12 (1): 25–34.

Ipsos MORI. 2010. *Experiment Earth? Report on a Public Dialogue on Geoengineering.* London, Eng.: Ipsos MORI, August.

Jepma, Catrinus J. 2008. "'Biosphere Carbon Stock Management: Addressing the Threat of Abrupt Climate Change in the Next Few Decades.' By Peter Read. An Editorial Comment." *Climatic Change* 87 (3-4): 343–46.

Keeling, C. D. 1960. "The Concentration and Isotopic Abundances of Carbon Dioxide in the Atmosphere," *Tellus* 12 (2): 200–03.

Keeling, C. D., S. C. Piper, R. B. Bacastow, M. Wahlen, T. P. Whorf, M. Heimann, and H. A. Meijer. 2001. *Exchanges of Atmospheric CO_2 and $^{13}CO_2$ with the Terrestrial Biosphere and Oceans from 1978 to 2000.* I. Global Aspects, SIO Reference Series, No. 01-06. San Diego, Calif.: Scripps Institution of Oceanography.

Keeling, R. F., S. C. Piper, A. F. Bollenbacher, and J. S. Walker. 2009. "Atmospheric CO_2 Records from Sites in the SIO Air Sampling Network." In *Trends: A Compendium of Data on Global Change.* Carbon Dioxide Information Analysis Center, Oak Ridge

National Laboratory. Oak Ridge, Tenn.: U.S. Department of Energy.

Keeling, Ralph. 2011. Personal correspondence to Ana Ivelisse Aviles, Senior General Engineer, U.S. Government Accountability Office, Washington, D.C., February 1.

Keith, David W. 2010. "Photophoretic Levitation of Engineered Aerosols for Geoengineering." *Proceedings of the National Academy of Sciences of the USA* 107 (38): 16428–31.

Kiehl, J. T., and Kevin E. Trenberth. 1997. "Earth's Annual Global Mean Energy Budget." *Bulletin of the American Meteorological Society* 78 (2): 197–208.

Kravitz, Ben, Alan Robock, Olivier Boucher, Hauke Schmidt, Karl E. Taylor, Georgiy Stenchikov, and Michael Schulz. 2011. "The Geoengineering Model Intercomparison Project (GeoMIP)." *Atmospheric Science Letters* 12 (2): 162–67.

Lacis, Andrew A., Gavin A. Schmidt, David Rind, and Reto A. Ruedy. 2010. "Atmospheric CO_2: Principal Control Knob Governing Earth's Temperature," *Science* 330 (6002): 356–59.

Laird, David A., Robert C. Brown, James E. Amonette, and Johannes Lehmann. 2009. "Review of the Pyrolysis Platform for Coproducing Bio-Oil and Biochar." *Biofuels, Bioproducts and Biorefining* 3 (5): 547–562.

Lane, Lee, Ken Caldiera, Robert Chatfield, and Stephanie Langhoff, eds. 2007. *Workshop Report on Managing Solar Radiation.* Report of a workshop jointly sponsored by NASA/Ames Research Center and the Carnegie Institution of Washington Department of Global Ecology held at Ames Research Center, Moffett Field, California, November 18-19, 2006. NASA/CP-2007-214558. Hanover, Md.: NASA Center for Aerospace Information, April.

Latham, John, Philip Rasch, Chih-Chieh Chen, Laura Kettles, Alan Gadian, Andrew Gettelman, Hugh Morrison, Keith Bower, and Tom Choularton. 2008. "Global Temperature Stabilization via Controlled Albedo Enhancement of Low-Level Maritime Clouds." *Philosophical Transactions of the Royal Society A* 366: 3969–87.

Lehmann, Johannes. 2007. "A Handful of Carbon," *Nature* 447 (7141): 143–44.

Lehmann, Johannes, John Gaunt, and Marco Rondon. 2006. "Bio-char Sequestration in Terrestrial Ecosystems—A Review." *Mitigation and Adaptation Strategies for Global Change* 11 (2): 403-427.

Leiserowitz, Anthony, Nicholas Smith, and Jennifer R. Marlon. 2010. *Americans' Knowledge of Climate Change.* Yale University. New Haven, Conn.: Yale Project on Climate Change Communication.

Lenton, T. M., and N. E. Vaughan. 2009. "The Radiative Forcing Potential of Different Climate Geoengineering Options." *Atmospheric Chemistry and Physics* 9 (15): 5539–61.

Levinson, Ronnen, and Hashem Akbari. 2010. "Potential Benefits of Cool Roofs on Commercial Buildings: Conserving Energy, Saving Money, and Reducing Emission of Greenhouse Gases and Air Pollutants." *Energy Efficiency* 3 (1): 53–109.

Lindzen, Richard S. 2010. "Global Warming: How to Approach the Science. (Climate Models and the Evidence?)." Testimony before the Subcommittee on Energy and the Environment,

Committee on Science and Technology, U.S. House of Representatives, 111th Cong. Washington, D.C., November 17.

Lindzen, Richard S., and Yong-Sang Choi. 2009. "On the Determination of Climate Feedbacks from ERBE Data." *Geophysical Research Letters* 36, L16705.

Long, Jane C. S. 2010. "Testimony to the House Science Committee: Geoengineering III Hearings." Testimony before the Committee on Science and Technology, U.S. House of Representatives, 111th Cong. Washington, D.C., March 18.

McCarl, Bruce A., Cordner Peacocke, Ray Chrisman, Chih-Chun Kung, and Ronald D. Sands. 2009. "Economics of Biochar Production, Utilization and Greenhouse Gas Offsets," pp. 341–56 in *Biochar for Environmental Management: Science and Technology,* Johannes Lehmann and Stephen Joseph, eds. London, Eng.: Earthscan.

McGuffie, K., and A. Henderson-Sellers. 2001. "Forty Years of Numerical Climate Modelling." *International Journal of Climatology* 21 (9): 1067–109.

McInnes, C. R. 2002. "Minimum Mass Solar Shield for Terrestrial Climate Control." *Journal of the British Interplanetary Society* 55: 307–11.

MacMinn, Christopher W., and Ruben Juanes. 2009. "A Mathematical Model of the Footprint of the CO_2 Plume during and after Injection in Deep Saline Aquifer Systems." Proceedings of the 9th International Conference on Greenhouse Gas Control Technologies (GHGT-9), Washington, D.C., November 16–20, 2008. *Energy Procedia* 1 (1): 3429–36.

Maibach, Edward, Connie Roser-Renouf, and Anthony Leiserowitz. 2009. *Global Warming's Six Americas 2009: An Audience Segmentation Analysis.* Fairfax, Va.: George Mason University, Center for Climate Change Communication, June 19.

Meehl, Gerald A., and Kathy Hibbard. 2007. *A Summary Report: A Strategy for Climate Change Stabilization Experiments with AOGCMs and ESMs.* WCRP Informal Report 3/2007, ICPO Pub. 112, and IGBP Report 57. Aspen Global Change Institute 2006 Session, Earth System Models: The Next Generation Aspen, Colorado, July 30–August 5, 2006 and Joint WGCM/ Aimes Steering Committee Meeting September 27, 2006.

Metz, Bert, Ogunlade Davidson, Heleen de Coninck, Manuela Loos, and Leo Meyer, eds. 2005. *IPCC Special Report on Carbon Dioxide Capture and Storage: Summary for Policymakers and Technical Summary.* Prepared by Working Group III of the Intergovernmental Panel on Climate Change. New York and Cambridge, Eng.: Cambridge University Press.

MITRE. 2011. *Risk Management Toolkit.* McLean, Va. www.mitre.org (Our Work, Mission Areas, Enterprise Systems Engineering, Systems Engineering Practice Office, Toolkits, Risk Management. Visited January 24.)

Morgan, M. Granger, and Katharine Ricke. 2010. *Cooling the Earth through Solar Radiation Management: The Need for Research and an Approach to Its Governance.* An Opinion Piece for International Risk Governance Council. Geneva, Switzerland: International Risk Governance Council.

Murphy, Daniel M. 2009. "Effect of Stratospheric Aerosols on Direct Sunlight and Implications for Concentrating Solar Power." *Environmental Science and Technology* 43 (8): 2784–86.

Murray, Brian C., Brent Sohngen, Allan J. Sommer, Brooks Depro, Kelly Jones, Bruce McCarl, Dhazn Gillig, Benjamin DeAngelo, and Kenneth Andrasko. 2005. *Greenhouse Gas Mitigation Potential in U.S. Forestry and Agriculture.* Office of Atmospheric Programs (6207J). EPA 430-R-05-006. Washington, D.C.: U.S. Environmental Protection Agency, November.

Nabuurs, Gert Jan, Omar Masera, Kenneth Andrasko, Pablo Benitez-Ponce, Rizaldi Boer, Michael Dutschke, Elnour Elsiddig, Justin Ford-Robertson, Peter Frumhoff, Timo Karjalainen, Olga Krankina, Werner A. Kurz, Mitsuo Matsumoto, Walter Oyhantcabal, N. H. Ravindranath, Maria José Sanz Sanchez, Xiaquan Zhang. 2007. "Forestry," pp. 543–84, in *Climate Change 2007: Mitigation,* B. Metz, O. R. Davidson, P. R. Bosch, R. Dave, and L. A. Meyer, eds. Contribution of Working Group III to the Fourth Assessment Report of the Intergovernmental Panel on Climate Change. New York and Cambridge, Eng.: Cambridge University Press.

NAS (National Academy of Sciences, Panel on Policy Implications of Greenhouse Warming). 1992. *Policy Implications of Greenhouse Warming: Mitigation, Adaptation, and the Science Base.* National Academy of Sciences, National Academy of Engineering, and Institute of Medicine. Washington, D.C.: National Academy Press.

Nisbet, Matthew C., and Teresa Myers. 2007. "The Polls—Trends: Twenty Years of Public Opinion about Global Warming." *Public Opinion Quarterly* 71 (3): 444–70.

Nolte, William L. 2004. *AFRL Hardware and Software Transition Readiness Level Calculator, Version 2.2.* Air Force Research Laboratory, Wright-Patterson Air Force Base, Ohio. www.acq.osd.mil/jctd/TRL/TRL%20Calc%20 Ver%202_2.xls

NRC (National Research Council). 2010a. *Advancing the Science of Climate Change.* Washington, D.C.: National Academies Press.

NRC (National Research Council). 2010b. *Ocean Acidification: A National Strategy to Meet the Challenges of a Changing Ocean.* Washington, D.C.: National Academies Press.

NRC (National Research Council). 2007. *Earth Science and Applications from Space: National Imperatives for the Next Decade and Beyond.* Washington, D.C.: National Academies Press.

Ohring, George, ed. 2007. *Achieving Satellite Instrument Calibration for Climate Change (ASIC³).* Report of a Workshop Organized by National Oceanic and Atmospheric Administration, National Institute of Standards and Technology, National Aeronautics and Space Administration, National Polar-Orbiting Operational Environmental Satellite System-Integrated Program Office, and Space Dynamics Laboratory of Utah State University, Lansdowne, Virginia, May 16–18, 2006. Washington, D.C.: National Oceanic and Atmospheric Administration.

Olson, Robert L. Forthcoming. *Geoengineering for Decision Makers.* Washington, D.C.: Woodrow Wilson International Center for Scholars.

Oruganti, YagnaDeepika, and Steven L. Bryant. 2009. "Pressure Build-Up during CO_2 Storage in Partially Confined Aquifers." Proceedings of the 9th International Conference on Greenhouse Gas Control Technologies (GHGT-9). Washington, D.C., November 16–20, 2008. *Energy Procedia* 1 (1): 3315–22.

OSTP (Office of Science and Technology Policy). 2010. *Achieving and Sustaining Earth Observations: A Preliminary Plan Based on a Strategic Assessment by the U.S. Group on Earth Observations.* Washington, D.C.: Office of Science and Technology Policy, September.

Parson, Edward A. 2006. "Reflections on Air Capture: The Political Economy of Active Intervention in the Global Environment: An Editorial Comment." *Climatic Change* 74 (1–3): 5–15.

Pearson, Jerome, John Oldson, and Eugene Levin. 2006. "Earth Rings for Planetary Environment Control." *Acta Astronautica* 58 (1): 44–57.

Pew (Pew Research Center for the People & the Press). 2010. "Public Remains of Two Minds on Energy Policy." *Congressional Connection.* Pew Research Center/National Journal Poll sponsored by the Society for Human Resource Management. Washington, D.C.: June 14.

Pielke Jr., Roger A. 2009. "An Idealized Assessment of the Economics of Air Capture of Carbon Dioxide in Mitigation Policy." *Environmental Science & Policy* 12 (3): 216–25.

Popp, David. 2006. "ENTICE-BR: The Effects of Backstop Technology R&D on Climate Policy Models." *Energy Economics* 28 (2): 188–222.

Popper, Steven W., Robert J. Lempert, and Steven C. Bankes. 2005. "Shaping the Future." *Scientific American* 292 (4): 66–71.

Ranjan, Manya. 2010. "Feasibility of Air Capture." M.S. thesis, Engineering Systems Division, Massachusetts Institute of Technology, Cambridge, Massachusetts.

Ranjan, Manya, and Howard J. Herzog. 2010. *Feasibility of Air Capture*. GHGT-10. *Energy Procedia*. Elsevier Use Only 00 (2010) 000-000. http://sequestration.mit.edu/pdf/ GHGT10_Ranjan.pdf

Rasch, Philip J., Paul J. Crutzen, and Danielle B. Coleman. 2008. "Exploring the Geoengineering of Climate Using Stratospheric Sulfate Aerosols: The Role of Particle Size." *Geophysical Research Letters* 35, L02809.

Rasch, Philip, John Latham, and Chih-Chieh (Jack) Chen. 2009. "Geoengineering by Cloud Seeding: Influence on Sea Ice and Climate System." *Environmental Research Letters* 4 (4):1-8.

Rasch, Philip J., Simone Tilmes, Richard P. Turco, Alan Robock, Luke Oman, Chih-Chieh (Jack) Chen, Georgiy L. Stenchikov, and Rolando R. Garcia. 2008. "An Overview of Geoengineering of Climate Using Stratospheric Sulphate Aerosols." *Philosophical Transactions of the Royal Society A* 366: 4007–37.

Rau, Greg H. 2009. "Geoengineering via Chemical Enhancement of Ocean CO_2 Uptake and Storage or Ignore Ocean Chemistry at Our Peril." Response to NAS request for input on geoengineering concepts. Lawrence Livermore National Laboratory, Livermore, California: June.

Rau, Greg H., Kevin G. Knauss, William H. Langer, and Ken Caldeira. 2007. "Reducing Energy-Related CO_2 Emissions Using Accelerated Weathering of Limestone." *Energy* 32 (8): 1471–77.

Read, Peter. 2008. "Biosphere Carbon Stock Management: Addressing the Threat of Abrupt Climate Change in the Next Few Decades: An Editorial Essay." *Climatic Change* 87 (3-4): 305–20.

Read, Peter, and Jonathan Lermit. 2005. "Bio-energy with Carbon Storage (BECS): A Sequential Decision Approach to the Threat of Abrupt Climate Change." *Energy* 30 (14): 2654–71.

Rejeski, David. 2010. "The Molecular Economy." *The Environmental Forum* 27 (1): 36–41.

Rejeski, David W. 2003. "S&T Challenges in the 21st Century: Strategy and Tempo," pp. 47–57 in *AAAS Science and Technology Policy Yearbook 2003*, Albert H. Teich, Stephen D. Nelson, Stephen J. Lita, and Amanda Hunt, eds. Washington, D.C.: American Association for the Advancement of Science.

Ridgwell, Andy, Joy S. Singarayer, Alistair M. Hetherington, and Paul J. Valdes. 2009. "Tackling Regional Climate Change by Leaf Albedo Bio-geoengineering." *Current Biology* 19 (2): 146–50.

Roberts, Kelli G., Brent A. Gloy, Stephen Joseph, Norman R. Scott, and Johannes Lehmann. 2010. "Life Cycle Assessment of Biochar Systems: Estimating the Energetic, Economic, and Climate Change Potential." *Environmental Science & Technology* 44 (2): 827–33.

Robock, Alan. 2000. "Volcanic Eruptions and Climate." *Reviews of Geophysics* 38 (2): 191–219.

Robock, Alan, Allison Marquardt, Ben Kravitz, and Georgiy Stenchikov. 2009. "Benefits, Risks, and Costs of Stratospheric Geoengineering." *Geophysical Research Letters* 36, L19703.

Robock, Alan, Luke Oman, and Georgiy L. Stenchikov. 2008. "Regional Climate Responses to Geoengineering with Tropical and Arctic SO_2 Injections." *Journal of Geophysical Research* 113, D16101.

Royal Society. 2009. *Geoengineering the Climate: Science, Governance and Uncertainty.* London, Eng.

Royal Society. 2001. *The Role of Land Carbon Sinks in Mitigating Global Climate Change.* Policy document 10/01. Prepared by the Royal Society Working Group on Land Carbon Sinks. London, Eng.: July.

Sabine, Christopher L., Richard A. Feely, Nicolas Gruber, Robert M. Key, Kitack Lee, John L. Bullister, Rik Wanninkhof, C. S. Wong, Douglas W. R. Wallace, Bronte Tilbrook, Frank J. Millero, Tsung-Hung Peng, Alexander Kozyr, Tsueno Ono, and Aida F. Rios. 2004. "The Oceanic Sink for Anthropogenic CO_2." *Science,* New Series 305 (5682): 367–71.

Salter, Stephen, Graham Sortino, and John Latham. 2008. "Sea-Going Hardware for the Cloud Albedo Method of Reversing Global Warming." *Philosophical Transactions of the Royal Society A,* 366: 3989–4006.

Sarmiento, Jorge L., and Nicolas Gruber. 2002. "Sinks for Anthropogenic Carbon." *Physics Today* 55 (8): 30–36.

Schuiling, R. D., and P. Krijgsman. 2006. "Enhanced Weathering: An Effective and Cheap Tool to Sequester CO_2." *Climatic Change* 74 (1–3): 349–54.

Secretariat of the Convention on Biological Diversity. 2009. *Scientific Synthesis of the Impacts of Ocean Fertilization on Marine Biodiversity.*

CBD Technical Series No. 45. Montreal, Quebec.

Shetty, Reshma P., Drew Endy, and Thomas F. Knight. 2008. "Engineering BioBrick Vectors from BioBrick Parts." *Journal of Biological Engineering* 2: 5.

Slingo, Julia, Kevin Bates, Nikos Nikiforakis, Matthew Piggott, Malcolm Roberts, Len Shaffrey, Ian Stevens, Pier Luigi Vidale, and Hilary Weller. 2009. "Developing the Next-Generation Climate System Models: Challenges and Achievements." *Philosophical Transactions of the Royal Society A* 367: 815–31.

Sohngen, Brent. 2009. *An Analysis of Forestry Carbon Sequestration as a Response to Climate Change.* Frederiksberg, Denmark: Copenhagen Consensus Center, Copenhagen Business School.

Sohngen, Brent, and Roger Sedjo. 2006. "Carbon Sequestration in Global Forests under Different Carbon Price Regimes," *The Energy Journal.* Multi-Greenhouse Gas Mitigation and Climate Policy Special Issue, Special Issue No. 3: 109–26.

Solomon, Susan. 1999. "Stratospheric Ozone Depletion: A Review of Concepts and History." *Reviews of Geophysics* 37 (3): 275–316.

Solomon, Susan, Dahe Qin, Martin Manning, Melinda Marquis, Kristen Averyt, Melinda M. B. Tignor, Henry LeRoy Miller, Jr., and Zhenlin Chen, eds. 2007. *Climate Change 2007: The Physical Science Basis.* Contribution of Working Group I to the Fourth Assessment Report of the Intergovernmental Panel on Climate Change. New York and Cambridge, Eng.: Cambridge University Press.

Solomon, Susan, Karen H. Rosenlof, Robert W. Portmann, John S. Daniel, Sean M. Davis,

Todd J. Sanford, and Gian-Kasper Plattner. 2010. "Contributions of Stratospheric Water Vapor to Decadal Changes in the Rate of Global Warming." *Science* 327 (5970): 1219–23.

Spencer, T., P. M. Guthrie, J. da Mosto, and C. A. Fletcher. 2005. "Introduction: Large-Scale Engineering Solutions to Storm Surge Flooding," pp. 241–43 in *Flooding and Environmental Challenges for Venice and Its Lagoon: State of Knowledge,* C. A. Fletcher and T. Spencer, eds. Cambridge, Eng.: Cambridge University Press.

Stauffer, Philip H., Ronald C. Surdam, Zunsheng Jiao, Terry A. Miller, and Ramsey D. Bentley. 2009. "Combining Geologic Data and Numerical Modeling to Improve Estimates of the CO_2 Sequestration Potential of the Rock Springs Uplift, Wyoming." Proceedings of the 9th International Conference on Greenhouse Gas Control Technologies (GHGT-9), Washington, D.C., November 16–20, 2008. *Energy Procedia* 1 (1): 2717–24.

Sullivan, Charlotte, Scott Frailey, James Damico, Joel Sminchak, Charles Gorecki, Kimberly Sams, Brian J. McPherson, David Borns, and Christine Doughty. 2011. *Risk Analysis and Simulation for Geologic Storage of CO_2.* DOE/NETL-2011/1459. Albany, Ore.: National Energy Technology Laboratory, March.

Taleb, Nassim Nicholas. 2007. *The Black Swan: The Impact of the Highly Improbable.* New York: Random House.

Tavoni, Massimo, Brent Sohngen, and Valentina Bosetti. 2007. "Forestry and the Carbon Market Response to Stabilize Climate," *Energy Policy* 35 (11): 5346–53.

Teller, E., L. Wood, and R. Hyde. 1997. *Global Warming and Ice Ages: I. Prospects for Physics-Based Modulation of Global Change.* UCRL-JC-128715. Preprint. Livermore, Calif.: Lawrence Livermore National Laboratory, August 15.

Tilmes, Simone, Rolando R. Garcia, Douglas E. Kinnison, Andrew Gettelman, and Philip J. Rasch. 2009. "Impact of Geoengineered Aerosols on the Troposphere and Stratosphere." *Journal of Geophysical Research* 114, D12305.

Tilmes, Simone, Rolf Müller, and Ross Salawitch. 2008. "The Sensitivity of Polar Ozone Depletion to Proposed Geoengineering Schemes." *Science Express,* April 24.

Trenberth, Kevin E., and Aiguo Dai. 2007. "Effects of Mount Pinatubo Volcanic Eruption on the Hydrological Cycle as an Analog of Geoengineering." *Geophysical Research Letters* 34, L15702.

Trenberth, Kevin E., and John T. Fasullo. 2010. "Tracking Earth's Energy." *Science* 328 (5976): 316–17.

Twomey, S. 1977. "The Influence of Pollution on the Shortwave Albedo of Clouds." *Journal of the Atmospheric Sciences* 34 (7): 1149–52.

U.S. House of Representatives. 2007. Committee on Appropriations. *Legislative Branch Appropriations Bill, 2008: Report to Accompany H.R. 2771.* H.R. Rep. No. 110-198. Washington, D.C.: U.S. Government Printing Office.

United States Senate. 2007. Committee on Appropriations, Legislative Branch Appropriations, 2008, *Report to Accompany S. 1686.* Rep. No. 110-89. Washington, D.C.: U.S. Government Printing Office.

Victor, David G. 2008. "On the Regulation of Geoengineering." *Oxford Review of Economic Policy* 24 (2): 322–36.

Washington, Warren M. 2006. "Computer Modeling the Twentieth- and Twenty-First-Century Climate." *Proceedings of the American Philosophical Society* 150 (3): 414–27.

Whitman, Thea, Sebastian M. Scholz, and Johannes Lehmann. 2010. "Biochar Projects for Mitigating Climate Change: An Investigation of Critical Methodology Issues for Carbon Accounting." *Carbon Management* 1 (1): 89–107.

Woolf, Dominic, James E. Amonette, F. Alayne Street-Perrott, Johannes Lehmann, and Stephen Joseph. 2010. "Sustainable Biochar to Mitigate Global Climate Change." *Nature Communications* 1 (56): 1–9.

Zeman, Frank. 2007. "Energy and Material Balance of CO_2 Capture from Ambient Air." *Environmental Science & Technology* 41 (21): 7558–63.

Zhang, Yan, Panagiotis Vouzis, and Nick Sahinidis. 2010. "Risk Assessment for CO_2 Geological Sequestration." Presented at the conference on Energy, Sustainability and Climate Change 2010, Center for Applied Optimization, University of Florida. Gainesville, Fla., February 26–28.

GAO contacts and staff acknowledgments

GAO contact

Timothy M. Persons, Ph.D., Chief Scientist, at (202) 512-6412 or personst@gao.gov

Other leadership for this project was provided by

Judith A. Droitcour, Ph.D., Assistant Director, Applied Research and Methods (ARM), and
Ana Ivelisse Aviles, Ph.D., Analyst-in-Charge and Senior General Engineer, also in ARM

Key contributors

Pille Anvelt, Visual Communications Analyst
Virginia A. Chanley, Ph.D., Senior Design Methodologist
Nirmal Chaudhary, Ph.D., Senior General Engineer
F. Kendall Childers, M.S., Senior Physical Scientist
Nancy J. Donovan, M.P.A., Senior Analyst
Gloria Hernandez-Saunders, Senior Information Technology Specialist
Eric M. Larson, Ph.D., Senior Analyst
Penny Pickett, Ph.D., Senior Communications Analyst
Ardith A. Spence, Ph.D., Senior Economist
Gregory H. Wilmoth, Ph.D., Assistant Director

In addition, **Juanita A. Aiken** provided engagement support, and **Leah Anderson** and
Katrina E. Pekar-Carpenter contributed as intern and student trainee. Other ARM
staff made contributions as did staff and specialists from GAO's Natural Resources and
Environment team and other GAO offices.

Related GAO products

Climate Change: A Coordinated Strategy Could Focus Federal Geoengineering Research and Inform Governance Efforts
GAO-10-903
Washington, D.C.: September 23, 2010

Coal Power Plants: Opportunities Exist for DOE to Provide Better Information on the Maturity of Key Technologies to Reduce Carbon Dioxide Emissions
GAO-10-675
Washington, D.C.: June 16, 2010

Climate Change: Preliminary Observations on Geoengineering Science, Federal Efforts, and Governance Issues. Testimony before the Committee on Science and Technology, U.S. House of Representatives, 111th Cong.
GAO-10-546T
Washington, D.C.: March 18, 2010

Climate Change Adaptation: Strategic Federal Planning Could Help Government Officials Make More Informed Decisions
GAO-10-113
Washington, D.C.: October 7, 2009

Climate Change: Expert Opinion on the Economics of Policy Options to Address Climate Change
GAO-08-605
Washington, D.C.: May 9, 2008

Global Warming: Limitations of General Circulation Models and Costs of Modeling Efforts
GAO/RCED-95-164
Washington, D.C.: July 13, 1995

Other GAO technology assessments

TECHNOLOGY ASSESSMENT:
Explosives Detection Technology to Protect Passenger Rail
GAO-10-898
Washington, D.C.: July 28, 2010

TECHNOLOGY ASSESSMENT:
Securing the Transport of Cargo Containers
GAO-06-68SU
Washington, D.C.: January 25, 2006
[Classification: For Official Use Only]

TECHNOLOGY ASSESSMENT:
Protecting Structures and Improving Communications during Wildland Fires
GAO-05-380
Washington, D.C.: April 26, 2005

TECHNOLOGY ASSESSMENT:
Cybersecurity for Critical Infrastructure Protection
GAO-04-321
Washington, D.C.: May 28, 2004

TECHNOLOGY ASSESSMENT:
Using Biometrics for Border Security
GAO-03-174
Washington, D.C.: November 14, 2002

GAO's mission

The Government Accountability Office, the audit, evaluation, and investigative arm of Congress, exists to support Congress in meeting its constitutional responsibilities and to help improve the performance and accountability of the federal government for the American people. GAO examines the use of public funds; evaluates federal programs and policies; and provides analyses, recommendations, and other assistance to help Congress make informed oversight, policy, and funding decisions. GAO's commitment to good government is reflected in its core values of accountability, integrity, and reliability.

To obtain copies of GAO reports and testimony

The fastest and easiest way to obtain copies of GAO documents at no cost is through GAO's website (www.gao.gov). Each weekday afternoon, GAO posts on its website newly released reports, testimony, and correspondence. To have GAO e-mail you a list of newly posted products, go to www.gao.gov and select "E-mail Updates."

To order by phone

The price of each GAO publication reflects GAO's actual cost of production and distribution and depends on the number of pages in the publication and whether the publication is printed in color or black and white. Pricing and ordering information is posted on GAO's website, www.gao.gov/ordering.htm.

Place orders by calling (202) 512-6000, toll free (866) 801-7077, or TDD (202) 512-2537.

Orders may be paid for with American Express, Discover Card, MasterCard, Visa, check, or money order. Call for additional information.

Congressional Relations contact

Ralph Dawn, Managing Director, at (202) 512-4400 or dawnr@gao.gov
U.S. Government Accountability Office, 441 G Street NW, Room 7125, Washington, DC 20548

Public Affairs contact

Chuck Young, Managing Director, at (202) 512-4800 or youngc1@gao.gov
U.S. Government Accountability Office, 441 G Street NW, Room 7149, Washington, DC 20548

To report fraud, waste, and abuse in federal programs

Web site: www.gao.gov/fraudnet/fraudnet.htm
E-mail: fraudnet@gao.gov
Automated answering system: (800) 424-5454 or (202) 512-7470